7-9-73

Evolutionary
Biology
of the
Primates

Evolutionary
Biology
of the
Primates

W. C. Osman Hill

Academic Press London and New York 1972

ACADEMIC PRESS INC. (LONDON) LTD.
24/28 Oval Road,
London NW1

United States Edition published by
ACADEMIC PRESS INC.
111 Fifth Avenue
New York, New York 10003

Library of Congress Catalog Card Number: 72-84449
ISBN: 0-12-528750-X

PRINTED IN GREAT BRITAIN BY
WILLMER BROTHERS LIMITED
BIRKENHEAD, CHESHIRE

Preface

The present volume owes its inception to an invitation which I received from Professor A. B. Chiarelli of the Department of Anthropology and Ethnology of the University of Turin to deliver a course of lectures suitable to the requirements of senior students on the subject of the biology of the Primates.

In the first place I should like to express my great indebtment to Professor Chiarelli for entrusting me with the organization of the syllabus for the proposed curriculum and secondly for enabling me to bring it to fruition in the shape of a text book.

My task has been materially assisted and rendered both easier and more pleasant by virtue of the help I have received from my many friends and professional colleagues, and particularly from the precursors in the well trodden field of Primatology that has so rapidly gained recognition in recent decades. I have specially to mention in this respect my friend, the late Sir Wilfrid Le Gros Clark, for the stimulus he has provided to myself and others in fostering the investigation of morphological and palaeontological problems in the Primate field. My own work has been rendered more meaningful as a direct result of the valuable texts and contributions to the anatomical and palaeontological journals.

Major problems have developed around the question of what subjects should be discussed and those which should be dismissed as of relatively lesser importance in the present context. Some selection becomes necessary in view of the rapid strides that have taken place in the last few years. I can but hope that my choice of subject matter has been a wise one and that I have done ample justice to the work of the pioneers yet, at the same time, incorporating significcant highlights culled from the researches of contemporary scholars.

It is a pleasure to acknowledge efforts in the field of illustration painstakingly undertaken by Mrs. C. O'Brien and Mrs. K. Watkins. Some of these are drawn specially from my originals, while others have been redrawn and modified from earlier works, including some of my own.

I owe sincere thanks to Edinburgh University Press for permission to reproduce from my "Primates, Comparative Anatomy and Taxonomy" (Vols 1–8) the following Figs., 16, 18, 24, 30, 68, 72, 75, 76, 78, 82, 85 and 86 and from Le Gros Clark's "Antecedents of Man" (1959) Figs. 19, 20, 21, 23, 26, 28, 29, 35, 38, 39 and 54.

From Academic Press I am duly grateful for permission to incorporate certain illustrations from Hopson's paper and one from Kreb's paper, taken from "Early Mammals" (Kermack and Kermack, eds.) Academic Press, 1971. To Heinemann's Medical Books I acknowledge permission to reproduce Figs. 3, 6, 13, 14, 22, 27, 31, 32, 33, 34 and 102 from "Man's Ancestry" (Hill, 1954).

My thanks are also due for illustrations borrowed and redrawn from certain Chapters in "Primatologia" (Hofer, Schultz and Starck, eds.) Karger, Basel, 1958, 1960; especially in contributions from Eckstein (Figs. 59, 60, 61, 62, 63, 64, 65 and 67, whence also Fig. 69 has been adapted from a paper of Wislocki) and Hill (Fig. 66). Figures 7, 12, 15 and 37 are reproduced from photographs by Miss Bessie Whitely and Fig. 25 from a photograph kindly put at my disposal by Rev. J. Cole. The photograph in Fig. 10 was supplied by Mr. M. Lister and is acknowledged to the Zoological Society of London. From the same source I have to acknowledge Fig. 9 which has been redrawn from a paper by Dr. P. M. Butler.

Istituto di Antropologia
   ed Etnografia,
Centro di Primatologia,
Università di Torino, Italia.

Royal College of Surgeons
       of England,
Lincoln's Inn Fields,
London, England.
August, 1972.

W. C. Osman Hill

TO MY WIFE

Contents

1

Primate Precursors

What is a Primate?

A Primate may be defined as any member of the mammalian order which is comprised by Man, Apes and Monkeys (Simiae) together with the so-called half-monkeys or Prosimians (Prosimii). Primates differ from other mammals mainly in negative features, rendering more exact definition difficult and controversial.

Mivart's (1875) definition however remains adequate for general purposes, except to those workers who wish to embrace the Tree-shrews (Tupaioidea) within the order:

> Unguiculate, claviculate placental mammals, with orbits encircled by bone; three kinds of teeth, at least at one time of life; brain always with a posterior lobe and a calcarine fissure; the innermost digit of at least one pair of extremities opposable; hallux with a flat nail, or none; a well-developed caecum; penis pendulous; testes scrotal; always two pectoral mammae.

Palaeontological studies of earliest (Eocene and Palaeocene) primates necessitates revision insofar as some of the above criteria of primate status (e.g. complete bony orbital rim) are there lacking, but the transitional character of these and of the Tree-shrews needs to be taken into consideration.

Precursors of the Primates

Structurally Primates are comparatively primitive and generalized mammals (cf. for example, Artiodactyla and Perissodactyla), so their origins must be sought among much lowlier types.

Mesozoic Mammals

True mammals first appeared in the Triassic phase of the Mesozoic, having emerged from one branch of the subclass of mammal-like reptiles (Theriodontia) that flourished during Triassic times.

The earliest known mammal (though not the first to be discovered) is represented by an almost complete cranium of *Tritylodon* from Triassic strata of South Africa. Studies by Watson (1942) have suggested that it was not yet fully mammalian, but merely an advanced theriodont reptile; but the decision is arbitrary and not of vital importance because, at this stage, the transition from reptilian to mammalian status is almost perfect.

There was, until the middle of the nineteenth century, a prejudice against accepting the existence of mammals prior to the Tertiary Epoch; albeit the first specimen, a minute mandible, was found in 1764 in the Stonesfield slate quarries (Jurassic) near Oxford, England. It remained unrecognized until 1828, when it was identified by Phillips and given the name *Amphilestes* by Owen (1859). Other mandibles had, meantime, been recovered from the Stonesfield slate and were studied by Broderip and Buckland convincing them of the error of the prevailing view. Their conclusion was supported by the French savant Cuvier, who went as far as to link these finds with the primitive marsupial opossum (*Didelphis*). Nevertheless, one of Broderip's specimens was given the noncommittal name *Amphitherium* by de Blainville (1838), who hoped thereby to dissociate it from the opossum and align it with reptiles. As Swinnerton (1960) remarks "It is a curious irony of advancing knowledge that this name fits in remarkably with the now established view that, while retaining traces of a reptilian ancestry, it was well on the way to being a full-blooded mammal".

These Jurassic mandibles differ from reptilian jaws in being comprised by a single bone (the dentary) which articulated directly with the cranium. They also bear a series of sharp teeth (up to 16 each side) differentiated into incisors, canine, premolars and molars. The molars presented all the main features of the typical mammalian molar and, in some forms, were highly specialized, precluding them thereby from being ancestral to later mammals.

Already during the Mesozoic at least three evolutionary lines are indicated:

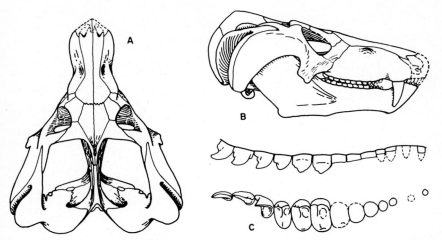

FIG. 1. *Diademodon*—a mammal-like reptile. A. Dorsal view of mature skull of *D. grossarthi*; skull length: 313 mm. B. Lateral view of mature (?) skull of *D. rhodesianus*; skull length: 236 mm. C. Lateral and crown views of right upper-postcanine dentition of *Diademodon* sp. in Berlin Museum; length of tooth row: 70 mm. Reproduced by courtesy of The Linnean Society of London.

1. The allotherians or multituberculates (e.g. *Plagiaulax, Stereognathus*; probably also *Tritylodon*). In all these the cheek teeth have more than three roots and their crowns bear three rows of six cusps arranged longitudinally. Some limb-bones have also been recovered and show characters intermediate between the theriodont reptiles and the bones of the duckbilled platypus (*Ornithorhynchus*), an egg-laying mammal of Australia, which retains several reptilian features in its skeleton.

2. The Triconodonta (e.g. *Amphilestes,* Middle Jurassic, England; *Priacodon,* Upper Jurassic, North America). These were small Mid- to Upper Jurassic mammals known chiefly from mandibles and teeth and some fragmentary cranial elements, which supply some information about the brain. Their cheek teeth bear three sharp conical cusps arranged in a single antero-posterior row – a pattern already shown in some theriodont reptiles. The triconodonts probably represented an evolutionary lineage that attained mammalian status independently, as they show no relationship to the multituberculates or to other Jurassic mammals, while intermediates are unknown. They could not, on their cusp pattern, have served as ancestors of any existing group.

3. The Trituberculata (sometimes referred to as Insectivora primi-tiva or Pantotheria) (e.g. *Amphitherium,* Mid-Jurassic, England; *Paurodon* and *Dryolestes,* Upper Jurassic, North America).

In these the cheek teeth present three sharp pointed cusps, arranged more or less in a triangle, the apex of which, in the mandibular teeth, is on the lateral side, the opposite being true of

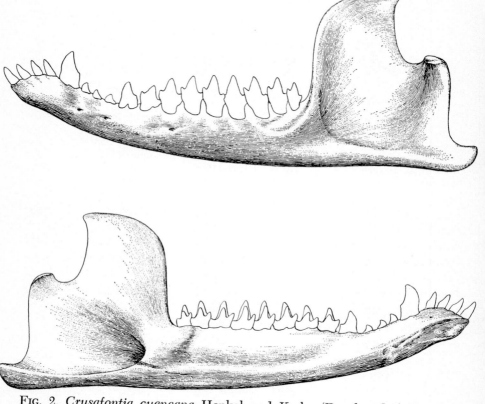

FIG. 2. *Crusafontia cuencana* Henkel and Krebs (Dryolestidae). Lower Cretaceous. Uña (prov. Cuenca, Spain). Restoration of a left lower jaw, combination of both finds from Uña. ×6·5. Upper: external view. Lower: internal view. Reproduced by courtesy of The Linnean Society of London.

the upper teeth. This is the basic pattern of modern mammalian molars in its simplest form. In the mandibular teeth also, a broad crushing surface (talonid basin) is borne on the crown behind the tubercles. The molar crowns are supported by three roots. This combination of features suggests that the pantotheres lay on or near the main line of mammalian descent, entitling them to be relegated to the Subclass Theria to which all modern mammals, with the exception of the egg-laying monotremes, *Ornithorhynchus*, and the echidnas belong. Hence the precursors of the Primates may safely be sought among the therians.

FIG. 3. Echidna, the Australian Spiny Anteater, a surviving monotreme.

CRITERIA OF THE SUBCLASS THERIA

Members of this vast assemblage all agree in the undermentioned features in which they contrast with the Subclass Prototheria, which is comprised solely by the monotremes or egg-laying mammals.

1. Sutures between many cranial bones do not early coalesce (this is a guarantee of long continued brain growth).
2. Vertebral centra are provided with plate-like epiphyses during the growing period; they finally fuse with the respective vertebral bodies.
3. There is no separate episternum; it is incorporated in the presternum.
4. Pectoral girdle elements coracoid and precoracoid are vestigial.
5. The pelvic girdle may or may not be provided with epipubic bones (marsupial bones).
6. The hallux (pedal digit I) is sometimes opposable to the other pedal digits, rendering the foot prehensile (e.g. opossums, phalangers).
7. The cerebral hemispheres of the brain may or may not be united by a large commissural band or corpus callosum.
8. A cloaca or common receptacle for the urinary, genital and alimentary discharges is not present. According to Gadow, this

statement is strictly applicable only to males; there exist some transitional states, notably among insectivores.

9. Oviducts of the two sides are united over part of their course and, in any event, each is differentiated into an anterior narrow segment (Fallopian tube), a broader thick-walled segment or uterus (often involved in the median fusion) and a terminal passage (vagina) leading to the exterior.

10. Mammary glands are provided with nipples.

11. Ova are holoblastic; early stages of development take place in the uterus. The embryo is there retained for a shorter or longer time and thereafter born alive through the vagina, along with its foetal membranes.

MARSUPIALS

There are two sections of the surviving Theria, both derived from the Trituberculates. One comprises the present-day marsupials (Metatheria or Marsupialia) represented by the kangaroo, wombats, koala etc. of the Australian region and by the primitive opossums (Didelphidae) of North and South America.

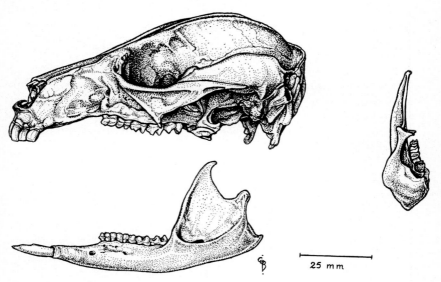

25 mm

FIG. 4. Cranium and mandible of a marsupial, the Rufous-bellied Wallaby (*Thylogale billardieri*). The inset shows the mandible from the rear, to indicate the peculiar marsupial tendency to inversion of the angular region,

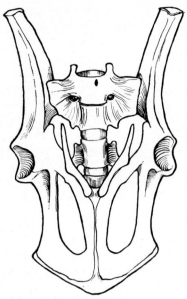

FIG. 5. Pelvic girdle of a marsupial,—kangaroo (*Macropus*) to show the epipubic bones. Redrawn from Wood Jones (1923).

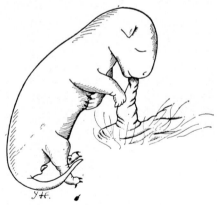

FIG. 6. Pouch-young of a Kangaroo (*Macropus*). Note the advanced state of development of the fore limbs compared with that of the hind pair, the closed eyes and the anchoring of the nipple to the pouch-wall.

In all marsupials the young are born in a very immature state and the placenta, when present, functions for a very brief period. After birth the embryos are transferred to the mammary area, where they become literally fixed, one to each nipple. The mammary area, except in some opossums, becomes outlined and protected by skin folds (mammary folds) which, in the more advanced forms, unite to produce a median pouch (marsupian) whose entrance may be directed anteriorly or posteriorly.

Marsupials are further characterized by:

1. The tympanic cavity in the skull is partly bounded by the alisphenoid, while the jugal (malar) provides a contribution to the glenoid cavity, where the dentary is articulated.
2. Epipubic bones present and well developed.
3. A single sphincter muscle encloses anal and urogenital apertures.
4. In males the scrotum is prepenial.
5. Corpus callosum lacking in brain.

From the Trituberculata (Insectivora primitiva) developed the true Insectivora and through the latter, directly or indirectly, all the remaining mammalian orders. This group-kinship is expressed taxonomically by placing all in the large infraclass Eutheria whose communal features, contrasting with the Metatheria, are:

1. Marsupial folds lacking; young born after prolonged uterine gestation, during which the foetal membranes produce a placenta that functions throughout intrauterine life.
2. In the skull the alisphenoid does not contribute to the wall of the tympanum, nor does the jugal assist in the formation of the glenoid (*Hyrax* and some rodents exceptional).
3. Epipubic bones lacking.
4. Anus and urogenital openings with separate sphincters.
5. Corpus callosum present.

2

The Transition to the Primates

Order Insectivora

Of the 23 orders of eutherians recognized at present, the only one that concerns us in connection with the antecedents of the Primates is the rather heterogeneous assemblage called Insectivora, represented today by such familiar types as shrews, moles and hedgehogs, but also embracing some lesser known though still important families from our standpoint.

All the insectivorous Eutheria agree in:

1. Rhinarium elongated, conical, sensitive and bedecked with numerous tactile vibrissae. Internally, a special sensory structure, Jacobson's organ, occurs on each side of the nasal septum.
2. Nocturnal mode of life; dependence on smell and tactile impressions for knowledge of the environment. Visual apparatus poorly developed or obsolete.
3. Clavicles present and plantigrade limbs terminating in pentadactyl clawed digits.
4. Dentition diphyodont, heterodont; the molars with high pointed cusps, and never less than two on each side in the mandible.
5. Brain simple, with smooth hemispheres.

A breakdown of the Insectivora into two main groups is possible in two different ways, each method having its supporters – as indicated in the appended scheme.

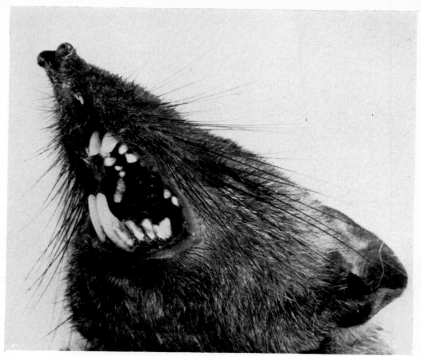

Fig. 7. Head of a White-toothed Shrew (*Crocidura*) photographed obliquely from below to show the characters of the rhinarium, vibrissae and type of dentition. $\times \frac{4}{1}$

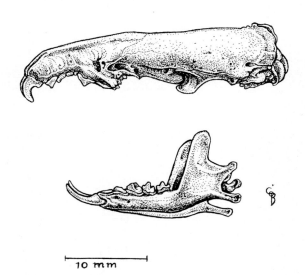

10 mm

Fig. 8. Skull, with mandible, of a typical lipotyphlous insectivore, an Indian Musk shrew (*Suncus caeruleus*). Note absence of bony ring surrounding orbit and the 180° relationship between facial and cranial axes.

Suborders according to Haeckel (1866)	Superfamilies of Simpson (1945)	Suborders according to Gill (1872)
1. Lipotyphla	Tenrecoidea Chrysochloroidea Erinaceoidea Soricoidea	A. Zalambdodonta
2. Menotyphla	Macroscelidoidea Tupaioidea	B. Dilambdodonta

From the point of view of the primatologist Haeckel's scheme is the more satisfactory, wherein the two families constituting the menotyphlous insectivores contrasted with the remainder or lipotyphlous branch in the possession of an ossified auditory bulla, in which the endotympanic is present and plays an important part; they also possess a caecum on the large intestine – lacking in all lipotyphlans.

The two menotyphlous families, Elephant shrews (Macroscelididae) and Tree shrews (Tupaiidae), though sharing many features also present some marked differences, which have been emphasized by Carlsson (1909, 1922) and by Le Gros Clark (1933) with the suggestion that the Macroscelididae be aligned with lipotyphlans and the Tupaiidae with the Prosimians, a course followed by many later students (e.g. Simpson (1945)). Others take the line that whatever the affinities of the tupaiids may be, the Elephant-shrews go with them. A detailed study of the osteology of Elephant-shrews has been made by Evans (1942), who found agreement with the Tree-shrews in 32 out of 40 characters said by Gregory (1910, 1913) and Carlsson (1922) to link *Tupaia* with the prosimians. Close relations between the two menotyphlan families and retention of Haeckel's grouping has been followed by several recent authorities (Grassé, 1955; McDowell, 1958).

This is not to say that the problem has been finally settled. van Valen (1965), on palaeontological grounds concludes that the tupaiid–primate relationship, though possible, is unlikely and Hill (1965), the embryologist, dispels one of the strongest arguments (placentation) that has been used to support the association, whilst studies by Martin (1966) on maternal behaviour also militate against a primate affiliation. On the contrary, Goodman's (1963) immunological studies of serum proteins indicate closer affinities with primates than with any other mammalian group.

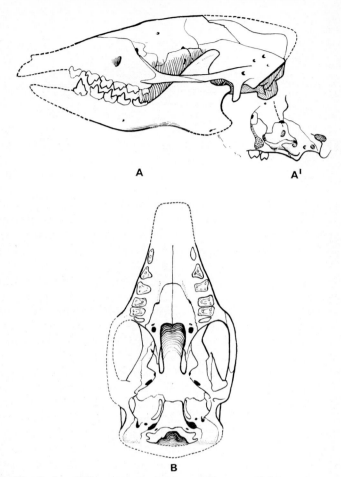

FIG. 9. Skull of the Oligocene insectivore *Ictops dakotensis,* in lateral view (A) and from the palatal aspect (B). In A¹ the details are shown of the bony contacts and foramina in the orbital region. Redrawn from P. M. Butler.

The Menotyphla differ from most lipotyphlans and resemble prosimians in the following ways (Butler, 1956):

1. Olfactory capsule shorter than brain, usually not extending between orbits (except partly in *Ptilocercus,* a primitive tree-shrew).
2. Cerebral hemispheres extending far forwards under frontals.
3. Postorbital constriction near middle of frontals.

FIG. 10. Four-toed Elephant-shrew (*Petrodromus*)—one of the family Macroscelidae. $\times \frac{2}{3}$. Reproduced by permission of Zoological Society of London.

4. Optic foramen and eyeballs large (adaptation to diurnal mode of life).
5. Postorbital processes and supraorbital crests well developed (Macroscelididae exceptional).
6. Maxilla not extending to medial orbital wall, hence no maxillo-frontal contact in orbit.
7. Orbital wing of palatine well developed.
8. Important part of orbito-sphenoid incorporated in brain-case.
9. Origin of m. temporalis not extending on to frontal.
10. Zygomatic arches complete; jugal well developed touching lachrymal.
11. Ectopterygoid process absent; origin of m. pterygoideus medialis extending to medial wall of orbit.
12. Coronoid process of mandible inclined backwards.

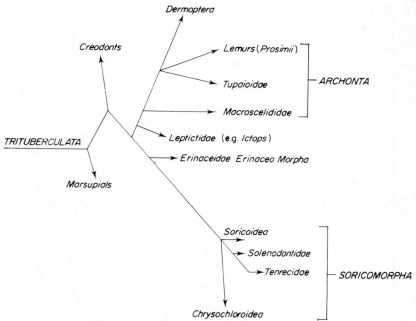

Fig. 11.

Butler (1956, p. 476) affirms that all these menotyphlan features are shared with the Dermoptera (Flying Lemurs) and the prosimian primates. He concludes that they represent a phyletic line altogether different from that leading to the true shrews and their allies (Soricomorpha), while the Erinaceomorpha (hedgehogs and their allies) occupy an intermediate position and, in general, are more primitive. He accordingly suggests the following arrangement (Fig. 11):

Superorder Archonta.
   Orders: 1. Macroscelidea
             2. Dermoptera
             3. Primates (including Tupaiidae)

Butler, from his studies on the Oligocene fossil *Ictops* of the primitive family Leptictidae, suggests the scheme in Fig. 11, modified in the light of McDowell's findings, to indicate the phylogeny of these animals.

The Tree-Shrews (Tupaiidae)

The annectant status of this family, between the Primates on the one hand and the lowlier mammals on the other and including even some marsupial affinities, is attested by the following criteria:
1. In the dentition, the upper molars are basically tritubercular with W-shaped crown pattern; the lower molars tuberculo-sectorial.
2. The alisphenoid exhibits a small tympanic process.
3. The organ of Jacobson is of marsupial type (Broom, 1915).
4. The scrotum prepenial, as in marsupials.
5. Placentation bidiscoid and endothelio-chorial (e.g. unlike that of lemurs, but resembling that of the higher Primates (Hill, 1965)).

Tree-shrews are small (average 18 cm. long) squirrel-like mammals with long rather pointed snouts and subequal pentadactyl limbs with all the digits bearing falcate, pointed claws. The tail is long and hairy (*Ptilocercus* excepted); mammae 2–6; dental formula $\frac{2.1.3.3.}{3.1.3.3.} = 38$; vertebral formula C.7, T. 13–14, L. 5–6, S.3, C. 22–31 (Cabrera, 1925).

They differ from Elephant-shrews in their smaller, simpler caecum and in the shorter foot. Moreover, postorbital processes are strongly developed, meeting the zygomata, the latter being perforate. Auditory bullae are more moderately developed.

BIOLOGY

The name Tree-shrew is somewhat of a misnomer, as most species are as much terrestrial as arboreal. The most important feature is

TABLE 1. Geological Time Scale Covering Mesozoic
and Subsequent Epochs.

Mill

HOLOZOIC	HOLOCENE								
	PLEISTOCENE								
CAINOZOIC = Tertiary	PLIOCENE								Man
	MIOCENE								
	OLIGOCENE								
	EOCENE								
	PALAEOCENE								
MESOZOIC = Secondary	CRETACEOUS					Marsupials	Eutheria	Primates	
	JURASSIC			Birds					
	TRIASSIC				Monotremes				
PALAEOZOIC			Mammal-like Reptiles						

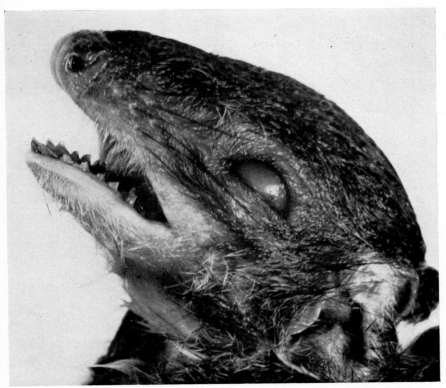

FIG. 12. Head of an Indian Tree-shrew (*Tupaia* or *Anathana ellioti*) to show the characters of the (strepsirhine) rhinarium, vibrissae and part of the lower dentition. $\times \frac{3}{1}$

their diurnal activity, which has lessened reliance on olfaction and induced the dependence on vision, with consequent change in the central nervous representation of these receptors.

Both hands are used in feeding. They are not, however, prehensile, the thumb being divergent but not opposable. In arboreal progression the hand is longitudinally oriented. Progression is entirely quadrupedal, with jerky scurrying motion like a rodent. Leaps across gaps of 4 feet are possible.

Exceedingly pugnacious, tree-shrews do not tolerate their own kind readily. They live singly or in pairs (Cantor, 1846). Communication is by vocal and visual signals (e.g. tail-flicking); forward lunges with open mouth is an aggressive display; presenting has been observed. Allogrooming, using partly procumbent lower in-

FIG. 13. *Tupaia (Tana) tana*. A typical Bornean tree-shrew.

cisors, and scratching with hands and feet is regular, but mutual grooming rare, though ♂ grooms ♀ at approach of oestrus. Scent marking of objects also occurs (Sprankel, 1961).

REPRODUCTION

Usually 2 (range 1–4) young produced after gestation of 41–50 days. Young born naked and blind. No limited breeding season. Oestrus cycle varies and may involve uterine bleeding.

DISTRIBUTION

India, Burma, Thailand, Indo-China, S. and W. China, Malay Peninsula and Islands, Borneo and Philippines.

Primates as an Evolutionary Series

The order is remarkable for the persistence, in spite of some extinctions, of living forms that illustrate several successive evolutionary steps in development of the order from tupaioids to man. This is not, however, to infer that all the stages represented constitute a linear phylogenetic series. The phylogeny shows evidence of branching lines as well as examples of parallel evolution and convergence.

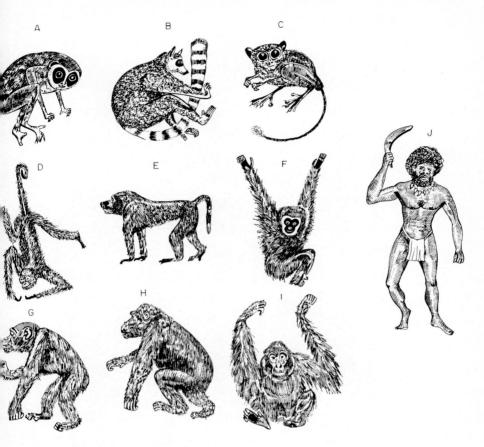

FIG. 14. Sketches of various primates illustrating a series comprised of some living species arranged in an ascending evolutionary order.

A. Slender loris (*Loris tardigradus*).
B. Ring-tailed lemur (*Lemur catta*).
C. Spectral tarsier (*Tarsius syrichta*).
D. Red-faced Spider monkey (*Ateles paniscus*).
E. Olive baboon (*Papio anubis*).
F. White-handed gibbon (*Hylobates lar*).
G. Chimpanzee (*Pan troglodytes*).
H. Gorilla (*Gorilla gorilla*).
I. Orang-utan (*Pongo pygmaeus*).
J. Man (Australian aboriginal (*Homo*)).

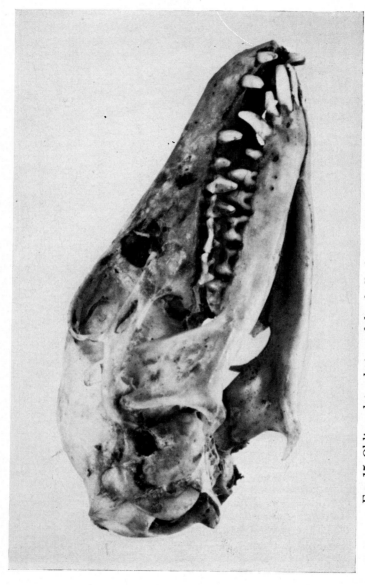

Fig. 15. Oblique lateral view of the skull of a tree-shrew (*Tupaia belangeri clarissa*) from Tenasserim, to show the orbit completely ringed by bone and the nature of the dentition. $\times \frac{3}{1}$

With this proviso, the following are the developmental evolutionary horizons represented:

1. *Menotyphlan Phase.*
2. *Prosimian Phase.*
   (Lemurs and their allies)
3. *Tarsioid Phase.*
   (*Tarsius* – sole survivor of a diversified group of Paleocene and Eocene fossils)
4. *Simian Phase.*
   (Arboreal or secondarily terrestrial quadrupedal monkeys)
5. *Anthropoid (Hominoid) Phase.*
   (Suberect brachiators)
6. *Hominid Phase.*
   (Erect bipedal terrestrial forms).

3

The Neurological Background of Primate Progress

How animals find their way about

Before proceeding to discuss these evolutionary horizons, it is essential to understand something of the neurological background as determined by the environmental adaptations developed in the successive phases. This digression is essential insofar as it is due to the increasing specialization of the brain that Primates owe their ascendency among the mammals, leading eventually to the emergence of man.

The brain is basically the repository for information received from the sense organs, which include olfactory, visual, auditory, tactile and other sources of external stimulation. The brain also serves to co-ordinate the messages received, to store, in memory, past experiences and, finally, to initiate the appropriate bodily reactions (either motor, glandular or both) on receipt of the information.

Receptor Organs

First it is necessary to examine the nature of the receptor organs responsible for receiving the stimuli from the environment.

As we have already seen, the earliest mammals were small and rather defenceless creatures that survived only by adopting a secretive, nocturnal existence, much as shrews and small rodents do today. They made very little use of visual systems, but lived in a world of smells, of tactile impressions and of high pitched sounds. Hence they depended upon their noses, their cutaneous sense organs and their auditory apparatus, all of which are specially adapted structurally and physiologically to this type of existence. Many of the adaptations are still to be met at the primate threshold.

OLFACTORY APPARATUS

The muzzle is, in these primitive forms, greatly elongated to form a snout or even a proboscis, upon and within which an organ complex is elaborated consisting of:

(a) The olfactory epithelium
(b) Jacobson's organ
(c) The rhinarium
(d) Several groups of enlarged bristly hairs (sinus hairs or vibrissae)

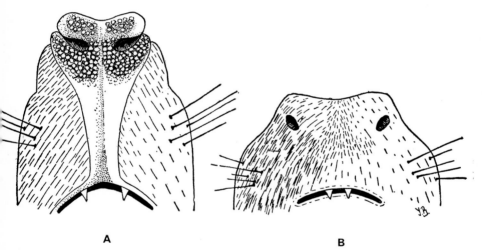

A   **B**

FIG. 16. Rhinarium of: A. a strepsirhine primate (lemur), compared with B. that of a haplorhine (Tarsius). Note the presence of sculpturing of the naked, moist areas in A.

(a) *Olfactory epithelium* is a specialized area of the lining of the nasal cavities, situated in the upper and posterior regions of the chambers. It contains receptor cells capable of chemical stimulation by minute particles carried in the respiratory stream and trapped in the nasal mucus. These specialized olfactory cells are true nerve cells connected by their axons, directly with the anterior part of the brain (olfactory lobe). This epithelium occurs in all mammals, but plays a bigger part in the lives of some (e.g. dogs) than others such as monkeys and men.

(b) *Jacobson's Organ* is a small structure situated on the lower front part of the nasal septum. Essentially it consists of a scroll of cartilage lying beneath the nasal mucosa and enclosing a cavity lined with an isolated patch of olfactory epithelium. The cavity communicates below with the mouth cavity by a small passage which perforates the bony palate just behind the upper incisor teeth. The lower opening of this passage (incisive canal) is a narrow slit lying in relation to a papillary thickening (papilla incisiva) on the fore part of the palatal mucous membrane. Physiologically, this pair of canals, by capillary attraction, is enabled to provide the chemically sensitive mucosa of Jacobson's organ with samples of saliva carrying potential food material for testing as regards palatability or otherwise. This organ has been inherited by the mammals from their reptilian forebears, but it has become vestigial in some of them, including man.

(c) *Rhinarium* is the name given to an area of specialized integument at the apex of the muzzle. It differs from normal skin in being hairless, smooth and moist, due to the presence of underlying glands. Its extent varies considerably, but when present it surrounds the external nares and occupies the region between them.

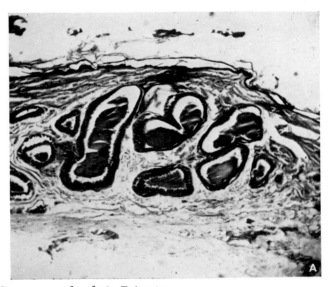

Fig. 17. Cutaneous glands in Primates
A. Vertical section of skin of the upper arm of *Loris tardigradus*, showing
 tubules of the brachial scent gland, filled with secretion. ×50.

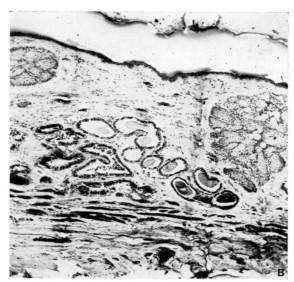

Fig. 17 B. Section of the scrotal skin of a marmoset (*Callithrix jacchus*) showing two types of glands. ×50.

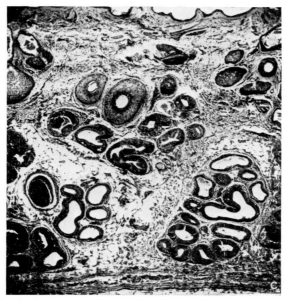

Fig. 17 C. Vertical section of the skin from the chest of a female Drill (*Mandrillus leucophaeus*) showing large coil-glands beneath the sectioned hair-follicles. ×40.

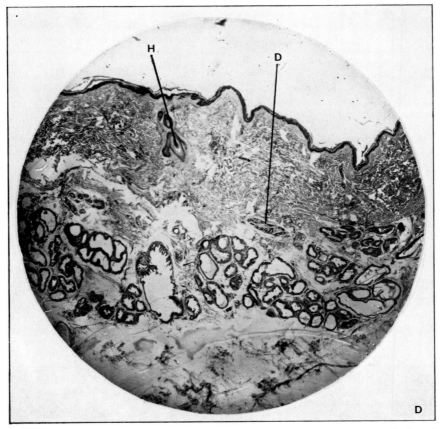

FIG. 17 D. Section of the axillary skin of a chimpanzee (*Pan troglodytes*) showing secreting coils and ducts of the axillary scent gland. ×17; H. hair follicle; D, duct.

From the internarial septum a downward extension occupies a tract of varying width that ends below by passing between the two halves of the upper lip (frequently this is bifid as in the rabbit, cat etc.) over the gum, thereby tethering the upper lip to the gum. Beyond the gum, the tract is brought into relation with the incisive papilla. Physiologically, therefore, the moisture of the rhinarium is enabled, by contact with external objects, food items, mates or other members of the creature's own kind, etc. to pick up minute particles or samples which are carried to Jacobson's organ for assessment. Various surface sculpturings serve to facilitate this process and also aid in tactile discrimination. For a comprehensive survey see Hill (1948).

(d) *Vibrissae or Sinus Hairs.* Finally in the head region generally, but especially about the muzzle and sometimes even elsewhere on the body (e.g. carpus) are developed groups of special hairs. All hairs are to some extent tactile organs inasmuch as contact of the hair shaft with an external object automatically distorts the follicle, within the walls of which are nerve terminals sensitive to the minute changes in pressure involved. Sinus hairs or vibrissae are lengthened and strengthened hairs and their follicles are modified by the development around them of large venous sinuses. Contact of a vibrissa with an external object involves greater distortion of the follicles and this sets up pressure waves in the blood sinus which magnify the effect on the surrounding nerve terminations. Long bristly vibrissae around the snout and on the interramal region project well beyond the body contours; hence they are able to give information about the size of a track, trail or crevice which a nocturnal, virtually sightless animal desires to enter. In this manner obstacles are avoided. In many lower primates groups of vibrissae are retained, while vestiges of some of them remain even among advanced forms (Fig. 18).

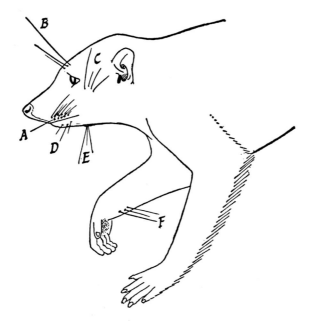

Fɪɢ. 18. Distribution of vibrissae on a typical strepsirhine primate. A. mystacial group; B. supraorbital group; C. genal group; D. mental group; E. interramal group and F. carpal group.

CUTANEOUS SCENT GLANDS

Before leaving the important subject of olfaction, the question of cutaneous glands must be briefly considered. Besides appreciating subtle differences in scents occurring in the environment, most mammals also produce scents as secretions from glands developed from their own skin. These cutaneous organs vary enormously in site, size and complexity, from small appendages to the hair follicles to large compact masses variously situated (see Schaffer (1940)). The functions of the secretions from these glands are numerous. In the first place, the secretion is imparted to the environment by bodily contact therewith, thus leaving a trail of the individual's movements. This enables individuals of a population to maintain contact. It also serves to bring males and females together at mating time. Scents therefore serve as a means of communication. Secretions are also deliberately applied to specific sites on the boundaries of an individual's territory (territory marking), thereby ensuring against over concentration of individuals. These phenomena are still to be found at work in the life of many primates including even, to a reduced extent, the anthropoid apes.

AUDITORY APPARATUS

The basic structural features of a receptor organ capable of responding to sounds and other vibrations has been inherited by the mammals from their reptilian ancestors. Many major modifications have, however, appeared in response to new requirements. Compared with the laboured and tardy reptilian responses, the early mammals, in order to survive, had of necessity to develop a higher metabolic rate. They needed to respond rapidly to environmental stimuli, so that early recognition of information derived from vibrations received from the substrate or from the air was essential. Hence we find the small primitive mammals like insectivores are restless hurrying creatures for ever on the alert.

The sensory cells which are responsive to vibratile stimuli are found in the walls of a cyst (otocyst) derived in the embryo from a downgrowth of a patch of skin (auditory placode) over the rear part of the head each side. The sensory cells send out processes or filaments which seek contact with an appropriate region of the brain stem.

In mammals, more especially the therians, elaboration from the simple cyst produces the structure known as the auditory labyrinth. In addition to the three semi-circular canals (already present in lower vertebrates), which are responsible for providing their owner with an appreciation of its orientation in space, there are areas capable of responding to vibrations from the substrate and a spirally coiled region – the cochlea (not present in reptiles) – whose sensory cells, constituting the organ of Corti, are arranged like the wires of a delicately tuned musical instrument. Many mammals are, additionally, able to appreciate ultrasonic vibrations, while some like the insectivorous bats, emit such vibrations so that their reflection from objects in the environment gives them information of obstacles to be avoided.

The labyrinth constitutes the internal ear which, during development, becomes incorporated within the petrous part of the temporal bone. Accessory structures comprising the middle ear and external ear, are devices serving to transmit sound and other vibrations from the environment to the buried internal ear.

The middle ear is an air-containing chamber derived embryologically from one of the pharyngeal pouches; it maintains a patent connection with the pharynx in the form of the Eustachian tube. The expanded upper end of this tube constitutes the tympanic cavity and is lined by an extension of the pharyngeal mucous membrane. One wall, the outer, is membranous (*tympanic membrane*). Around the periphery of the membranous area a supporting ring of bone is developed (tympanic ring, os ectotympanicum). Above, this ring generally gains attachment to the temporal bone. A bony floor (tympanic plate) for the tympanic cavity is developed inferiorly in all but the most primitive insectivores. It is generally derived as an outgrowth from the petrosal, but in insectivores contributions from the basisphenoid and alisphenoid are also made. Often in small nocturnal mammals the bony floor is expanded into a smooth rounded bulb (tympanic bulla) which acts as a resonating chamber, magnifying vibrations received by bone conduction from the substrate. In some noctural primates (lorises) a contribution is made by the os ectotympanicum. In the true lemurs on the other hand, as well as in New World monkeys, the expanding bulla has outstripped the tympanic ring enclosing it almost completely from below. The bulla is also subject to internal modifications, e.g., subdivision by bony septa.

The inner or medial wall of the tympanic cavity is supported by the petrous bone in the region where it encloses the labyrinth. Two small openings, filled with membrane (fenestra rotunda and fenestra ovalis) permit vibrations crossing the tympanic cavity to be transmitted to the fluid (perilymph) which is contained in the bony labyrinth and so, by pressure waves, to stimulate the sensory cells in the walls of the membranous labyrinth. The transmission is further facilitated by the presence of a chain of small bones (auditory ossicles) which cross the chamber transversely. The outermost ossicle or malleus is attached to the inner surface of the tympanic membrane. It is moveably articulated with the next or incus and this, in turn with the stirrup-shaped stapes, whose footplate is embedded in the membrane of the fenestra ovalis. That of the fenestra rotunda serving to buffer the pressure transmitted through the perilymph. These auditory ossicles thus function in the same way as a single bony strut or columella in the middle ear of reptiles and birds.

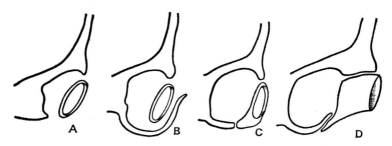

FIG. 19. Diagrams to illustrate the structural variations of the tympanic ring (os ectotympanicum) in mammals. The condition in: A. primitive mammals; B. lemuroids; C. lorisoids and platyrrhine monkeys and D. modern tarsioids and Old World monkeys.

The auditory ossicles are derived from some of the lost elements of the reptilian mandible that have become pinched off during phylogeny by the growth of the tympanic region of the skull.

The external ear is comprised of a tubular external auditory meatus around the superficial opening from which is developed, as a distinctly mammalian peculiarity, a funnel shaped cartilage-supported extension, known as the pinna or auricle.

The deeper part of the meatus may be supported by bone derived from an extension from the tympanic ring, e.g. in lorises, *Tarsius* and most higher forms. The pinna is of variable development. It serves the purpose of catching sound waves, channelling them into the meatus and is enabled to do this more effectively by the development of extrinsic muscles which endow it with the power of swivelling in various directions, so that the direction of the source of sound may be recognized. Diminution of the size and mobility of the pinna in the higher primates and man might lead to the inference that auditory acuity is of lesser grade than in lowlier forms with large mobile membranous ears. This is doubtless true in some measure, but it must be remembered that these large membranous ears, liberally supplied with blood vessels, also function as part of the temperature regulating mechanism, facilitating the loss of body heat, especially in small creatures with a high metabolic rate and in desert or tropical conditions. The relationship between size of pinna and general body size, in this connection, is well illustrated by the contrast between the small ears of the slow moving lorises and pottos and the large, membranous organs of the actively leaping galagos. All these related forms are nocturnal and inhabit similar tropical rain-forest environments. In the case of the tupaioids, the contrast between the large membranous ears of the primitive *Ptilocercus* (pen-tailed tree-shrews) and the reduced organs of *Tupaia*, where the ears are curiously reminiscent of their human counterparts, is explained by the fact that *Ptilocercus* is nocturnal, while *Tupaia* is diurnal, though both are highly active arboreal climbers (Le Gros Clark, 1959).

VISUAL APPARATUS

Although vision has little or no functional significance to the primitive nocturnal mammal in finding its way about, it becomes of great value when diurnal activity is adopted, e.g. in all tupaioids, except *Ptilocercus*. Curiously enough among some nocturnal primates, notably *Tarsius*, the lorises, some small nocturnal lemurs and the platyrrhine Night Monkey (*Aotes*), the eyeballs are strikingly enlarged. Increasing dominance of vision over other sense organs played a large part in the evolution of the higher primates.

The eyeball with its accessory structures, is a remarkable organ complex with parts derived from all the tissue-forming elements in the embryo. The sensory membrane of the eye, containing the cells capable of stimulation by light (photo-receptors), is the *retina* which also embraces layers of nerve cells and fibres which, by way of the optic nerve, reach the brain. The retina constitutes the innermost of the three coats of which the wall of the globe consists. Embryologically it is an outgrowth from the brain, the optic nerve being the stalk representing the direction taken by the outgrowth. The outer and middle coats of the globe, including accessory structures such as the lens, are arranged in relation to the retina by differentiation from neighbouring embryonic tissue, apparently by some evocative influence of the retina itself.

Photo-receptors are of two kinds called rods and cones. Rods have a high sensitivity, functioning at low levels of luminosity, but

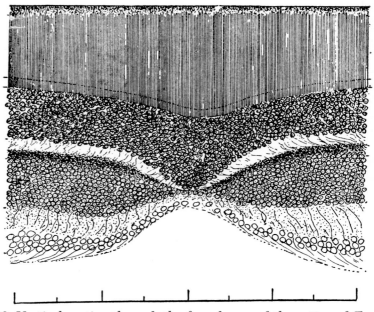

FIG. 20. Vertical section through the foveal area of the retina of *Tarsius*. The ganglion cells lie nearest to the scale, each of which marks a distance of 50 μ. From Le Gros Clark, after Polyak.

they are weakly discriminative. In nocturnal mammals they alone are present, for they come into action in twilight (or scotopic) vision. Rods are extensively distributed even in the peripheral parts of the retina, whence they are particularly sensitive to the movements of external objects in the peripheral parts of the visual field.

Cones, on the contrary, serve the needs of the photopic vision, since they are responsive to light rays of high intensity. They are capable of a high degree of discrimination of spatial relationships, especially in primates, where the optic axes of the two globes are arranged more or less parallel, so that the resulting overlap in the two fields permits stereoscopic (i.e. three-dimensional rather than panoramic) vision. Cones also are involved in the appreciation of texture and colour.

Illustrative contrasts may be cited. *Ptilocercus*, being crepuscular in habit, has a pure rod retina, while in diurnal tupaioids cones are present. Rods are also the sole sensory elements in the lorises and in such nocturnal lemurs as *Cheirogaleus*, but in *Lemur*, all species of which genus are active in daylight, large numbers of cones are present.

Examination of the retina of a series of primates shows that steady evolutionary advance has been effected. Concentration of cones in the central area of the retina becomes gradually evident. The central area also, in the higher forms, undergoes local differentiation to form a yellow spot (macula lutea) which contains no rods whatever. In the middle of the macula is an attenuated area, forming a small depression or fovea. In this, all the constituent layers of the retina have disappeared, except for the cones. This area is therefore the region of greatest visual acuity, for the light rays here reach the cones without the necessity of passing through other retinal layers or interference from blood vessels. A fovea is not present in lower primates such as lorises, galagos and lemurs, but the vessels are fewer and more delicate in the central area, as is also the case in *Tupaia*. In the strange nocturnal *Tarsius*, in spite of its purely rod retina, a well-developed yellow-pigmented macula with a small fovea has been demonstrated opthalmoscopically and confirmed histologically (Polyak, 1957). With this exception, a macula is otherwise confined to the monkeys, apes and man. A purely rod retina also occurs in the eye of the nocturnal New World monkey *Aotes*, which like *Tarsius*, bears a fovea. It has been suggested that the fovea in these examples is a vestige persisting in a primate of secondarily acquired nocturnal habit, descended from diurnal ancestors possessing cone retinas. This may also account for the fovea in *Tarsius* on the assumption that the Eocene ancestral tarsioids, or at least some of them, were diurnal. From the size of the orbits of such early tarsioids as *Tetonius* and *Pseudoloris*, however, these resembled the surviving *Tarsius* in being nocturnal.

The foveal depression is not so profound in the retina of marmosets as in monkeys generally, insofar as the retinal stratum known as the outer nuclear layer (i.e. the layer of the nuclei of the photoreceptor cells) stretches across it masking the cones from direct exposure to the light stimuli, thus a condition closely resembling that in *Tarsius* prevails.

At the other extreme, the macular region of the retina of the mangabey monkeys (*Cercocebus*) is more highly differentiated as judged by the density of photoreceptors, the perfection of the foveal excavation and the proportion of cones to rods, than in the human retina. Functionally, this is correlated with experimental evidence of a high degree of colour perception.

The rise and evolutionary advancement of the primates has been greatly assisted by the acquisition of stereoscopic vision. This was attained, as already mentioned, by a shift in the optic axes of the two eyeballs. In mammals such as horses and cattle, the eyes are situated on the sides of the head, the optic axes being almost transversely aligned. There is consequently little or no overlap in the visual fields of the two sides, each retina receiving a different picture. Vision is therefore referred to as panoramic. Among primates, the optic axes tend to become less and less divergent as the eyes shift their direction more forwards. The degree to which this has occurred shows wide variations among the prosimians, but in the higher primates the shift or rotation of the optic axes occurs during the embryological development of each individual, so that the eyes come to look directly forward and the axes of the two sides are parallel; the plane of the orbital opening in the skull follows this shift. In consequence there is an almost complete overlap in the visual fields, permitting any point in the total field to be focussed simultaneously on corresponding points of both retinae. The shift has been facilitated by and was indeed impossible until there had been a recession of the muzzle concomitant with reduced dominance of the olfactory sense.

Elliott Smith (1927) emphasized that evolution of a macula lutea rendered possible a better appreciation of the details, the texture and the colour of objects in the environment and an improved discrimination of their exact size, shape and position in space. These discriminations were further advanced with the advent of stereoscopic vision, as well as facilitating the perfection of tactile skill and manual dexterity through the more precise control, via the brain centres, of the movements of the hands and digits. Not by the advances in the visual apparatus alone, however, is the progression of primate evolution assured, but by this combined with certain advances in the structure and functions of the hands and feet, particularly the retention of prehensile digits and the elaboration of the tactile pads on the palms and soles. Details of the latter will be discussed hereafter.

4

Evolution of the Primate Brain

Information from the receptor cells in the various sense organs is relayed in the form of nerve impulses to the brain, where they are received, analysed, co-ordinated with information from other sense organs and compared with previous experiences. Finally the results of the analysis are rapidly processed and the appropriate muscular or other reactions instigated through other elements in the central nervous system.

The bulk of these cerebral functions are carried out, directly or indirectly, by the cerebral hemispheres, particularly the surface grey matter (cerebral cortex or pallium). The expansion and increased structural complexity of the cortex throughout the Primate series serves as an index of the evolutionary status of the owners of the brains.

Cerebral hemispheres originate as paired hollow outgrowths from the forebrain vesicle of the embryo and fundamentally are concerned with the olfactory reception. The forebrain vesicle also gives rise to a small hollow outgrowth on each side that develops into the sensory receptor field (*retina*) that becomes enclosed within the eyeball. Its connection with the brain persists as the optic nerve which, however, loses its hollow character.

The pallium, at first smooth, becomes, to a varying degree, folded due to the increase in the number of its component nerve cells and the increasing complexity of their fibre connexions and interconnexions. Hence, its hitherto smooth surface becomes marked by depressions (sulci) and elevations (gyri or convolutions). All stages between the simple smooth hemisphere and a highly convoluted one are represented among living Primates.

The Rise of the Neopallium

As already indicated, the pallium arose ostensibly to serve as a central reservoir for the receipt of olfactory impressions, for smell is perhaps, as we have seen, the most important sense to the primitive nocturnal mammal. The sensory cells that act as immediate smell receptors are found amongst supporting elements in the lining of the nasal chambers. These cells send their conducting axons to the extreme anterior part of the forebrain vesicle and adjacent parts of the outgrowing hemisphere, where they end around the receiving centres or nuclei. The rostral end of each hemisphere comes to form a bulbous structure (olfactory lobe) constricted off from the rest of the hemisphere by a narrow neck or tract (olfactory tract). This tract carries a further relay of fibres connecting cells in the olfactory lobe with others in the hemisphere proper, the whole of which, in its earliest form (as seen in primitive vertebrates), is concerned with further sorting and co-ordinating of olfactory impulses.

With the advent of newer information derived from tactile, visual and auditory senses – all demanding central representation – a new area of pallium is evolved, marked off morphologically from the ancient field (archipallium) by a sulcus (fissura rhinalis) and designated neopallium. It is to the elaboration and vast expanse of this region that the enlargement of the hemisphere in higher mammals is due. The success of the mammals is due to the result of the effect of the neopallium on their behaviour, for it enables them to profit, to a greater degree than in any other group of vertebrates, by individual experience. The neopallium renders this possible by its power of co-ordinating experiences from smell with those arising from other sense organs (especially those of sight, hearing and touch). In terms of psychology, the neopallium forms the physical basis which permits intelligence to override mere instinct, which is the dominant feature of reptilian and even most avian behaviour.

When it first appears, the neopallium occupies the dorsal or medial walls of the growing hemisphere, but among the mammals its expansion is so rapid, that it eventually overtops the archipallium, pushing it downwards and inwards until it is completely overshadowed by the new growth. Meanwhile the neopallium itself undergoes differentiation of its structural elements whereby, through histological and other methods, specific areas (*Brodmann's areas*) can be recognized and correlated with particular sense organs (or with areas responsible for the initiation of appropriate motor reactions).

The simplest picture of brain structure at this level is shown by

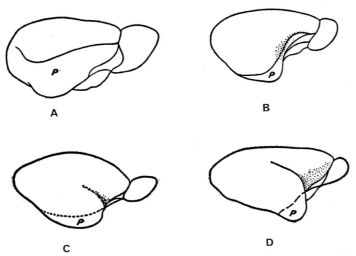

Fig. 21. Cerebral hemisphere, from the right side in a primitive lipo-typhlous insectivore, *Echinosorex* (A) compared with (B) that of *Tupaia*; (C) *Adapis*, from a plaster cast, and (D) *Microcebus*. p = pyriform lobe. After Le Gros Clark.

the brain of the Elephant Shrew (*Macroscelides*) – a menotyphlan insectivore. In side view, the large olfactory bulb and peduncle, with the adjacent part of the cerebral hemisphere (*pyriform lobe*) account for approximately half of the total pallium. The rhinal fissure runs more or less horizontally across the surface, marking off the dorsal moiety of the hemisphere (neopallium). At least 5 structurally differentiated areas in the latter can be detected by histological techniques; they are:

1. An insular or pararhinal area.
2. A general sensory area.
3. An acoustic or temporal area.
4. A motor area.
5. A frontal association area.

The general sensory area receives what are known as kinaesthetic or somatic sensory impulses, which are brought in chiefly via the spinal cord and brain stem. They are derived from muscles, tendons, bones and joints and serve to give information of the orientation of the bodily parts in space and in relation to each other. This is an ancient system whose higher control was formerly centred in a large basal nucleus, the *thalamus*, situated at the top of the brain stem. Higher representation in the neopallium is a more recent development found in mammals enabling these impulses to be more critically analysed and more readily co-ordinated with the other sensory systems.

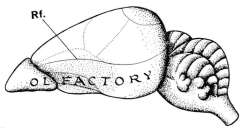

Fig. 22. Brain of an Elephant Shrew (*Macroscelides rufescens*) from the left side to indicate the areas devoted to olfactory reception. Rf : rhinal fissure.

The motor area lies adjacent to the preceding area, so that its nerve cells, responsible for initiating the stimuli for muscular activity, are arranged *pari passu* with their receptor counterparts.

Association areas are groups of cortical nerve cells that are brought into relation with the other receptor and motor fields by means of inter-communicating fibres. Their duty is clearly of a co-ordinating and assessing nature.

A marked advance is seen on comparison of the brain of the basically nocturnal *Macroscelides* with that of a diurnal tree shrew (*Tupaia*). In general, the brain of *Tupaia* is relatively larger as shown by the following data:

	Body weight	Brain weight	Index
Tupaia javanica	80	2·3	28·7
Talpa (Mole)	77	1·5	20
Microcebus murinus (Mouse lemur)	62	1·9	30

In *Tupaia* the olfactory lobe is relatively reduced and flattened, while the pyriform cortex has been displaced downwards and medially by the expanding neopallium, so that little of it can be seen from the lateral side. The corpus callosum, the great commissural tract which unites the neopallial cortex – of one side with corresponding parts of the other – has elongated and straightened to the degree exhibited by primitive lemurs such as *Microcebus*. The anterior commissure, which similarly serves the archipallium, on the contrary is small.

The neopallium has expanded chiefly in the occipital and temporal regions, but the front lobe remains much as in *Macroscelides*. The temporal lobe projects downwards to form a distinct "pole" in front of which the hemisphere is excavated (Sylvian fossa) to accommodate the bony roof of the enlarged orbit. These modifications have affected the course of the rhinal fissure – no longer straight and horizontal, but S-shaped, due to downward deflection of its hinder part.

Histological elaboration of the neopallium and the cortical areas have become as clearly circumscribed as in *Microcebus*. A remarkably well developed visual area occurs round the occipital pole on both aspects of the hemisphere, extending on the medial face almost to the corpus callosum.

The frontal cortex and adjacent association area remain much as in *Macroscelides*. In ordinary Insectivores, increase in body size (e.g. *Echinosorex*) is accompanied by increased brain size, especially affecting the pyriform lobes. However, in tupaioids, it is as in Primates, the increase in brain affects the neopallium at the expense of the pyriform lobe. Another part of the brain, the thalamus – a great sensory station at the base of the forebrain – is also affected in this enlargement. Within the thalamus is a relay station (lateral geniculate nucleus) in the fibre systems transmitting visual impressions to the neopallial cortex.

In *Tupaia* this body is not only large, but its cells are arranged in an imperfect multilaminar formation, foreshadowing the condition in true Primates (see also Hassler (1966)).

A transitional stage is represented in the brain of the Pen-tailed Tree-shrew (*Ptilocercus*) which, unlike *Tupaia*, is crepuscular in its activities rather than diurnal. Its brain shows the olfactory parts more like those in *Macroscelides* and the visual apparatus poorly represented (Le Gros Clark, 1959).

5

The Emergence of the Primates

The rise of the Primates from some relatively primitive insectivore ancestral group appears to have depended upon the perfection of adaptation to a diurnal arboreal existence, wherein the dependence upon the visual, auditory and tactile stimuli dominated the old world of odours.

In adopting an arboreal life limbs are no longer used solely or mainly for running or even jumping – the simplest forms of bodily progression – but the hands and feet become modified in several ways, especially giving them the power of grasping branches so as to give security and stability in the new environment, while the limbs, especially the forelimbs, are used to lift and draw the body forwards. Digits become elongated and more independent of each other; they are adapted to hook around branches, and the first digit (pollex on hand, hallux on foot) tends to diverge to a greater degree from the others than these are from each other. Thus the prehensile capacity is initiated. Thereby the digits take over the functions previously subserved by claws (falculae). Claws, as such, are therefore no longer necessary; they are reduced and converted into flattened nails (ungulae), for they fashion themselves in accordance with the shape of the terminal phalanges whose tips are expanded into a disc-like form in co-relation with the physiological action of pressure on the arboreal substrate (Winge, 1924).

Another highly important and significant modification of the hands and feet concerns their hairless palmar aspects. Certain elevations of the naked palmar and plantar integument, the so called touch pads (torulae tactiles) are inherited in their entirety from the insectivore ancestor. These are arranged in three groups: proximal,

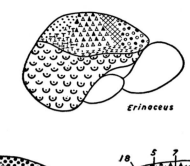

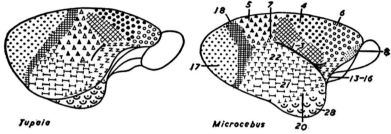

FIG. 23. Cerebral hemisphere, from the right, in *Erinaceus* (Hedge-hog); *Tupaia* and *Microcebus*, to demonstrate the cortical areas of Brodmann (1904).

Key to areas:

1–3	General sensory
4	General motor
5 and 7	Parietal
6, 8, 9	Frontal
13–16	Insular
17 and 18	Visual
19	Occipital
20–22	Temporal
28	Pyriform

After Le Gros Clark.

intermediate and distal. The proximal set consists of two large pads lying along the medial and lateral sides of the palm (and sole). The one on the thumb side is termed thenar pad; the other, sometimes divided into a proximal and a distal portion, is termed hypothemar. Between them the central part of the palm is depressed.

The intermediate pads are termed interdigital pads; ideally they are 4 in number and are borne on the distal part of the palm, beyond the previously named pads, one opposite each interdigital cleft. The first one is frequently fused with the thenar pad. Finally, the apical pads are carried on the expanded bulbous tips of each of the digits.

Primates are distinguished by further elaboration of the integument of the tactile pads. Summits of the pads show a tendency to the development of alternate ridges (papillary ridges) and grooves. The ridges are formed chiefly by elevations in the dermis to which the overlying epidermis conforms. Ridges are perforated by the ducts of sweat glands whose mouths appear in rows on the crests of the ridges. Ridges are also endowed with sensory nerve endings. Primarily the primitively ridged areas are the summits of the pads where they form a pattern of concentric circles around a short central ridge or group of such. This may be the full complement in the lowest forms, apart from systems which adorn the expanded apical pads on the fingers and toes. In this situation a longish central ridge (*fasciculus principalis*), aligned longitudinally, is flanked by parallel ridges which distally arch around the distal end of the fasciculus principalis.

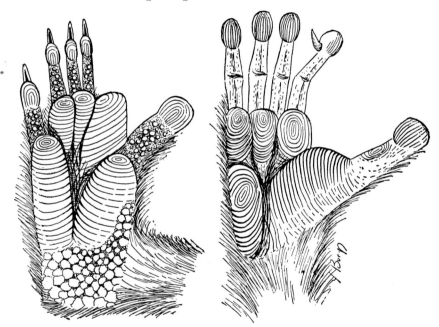

FIG. 24. Palmar aspect of the right foot of (left) a Virginian Opossum (*Didelphis marsupialis*) compared with (right) that of a Needle-clawed Galago (*Euoticus*).

Advances on this arrangement occur in various directions and by various methods. From the torulae, ridges invade the intervening smooth skin until the whole palm and sole become ridged. In *Lemur* this takes on an intermediate condition insofar as ridged islets of epithelium are developed on the areas between the principal pads. These show a tendency to fusion in smaller or larger islands, but there is always some unridged skin. In lorises, on the other hand, the whole surface is ridged by gradual spread from the torulae. In monkeys, ridges, either transversely disposed or arranged in V-formation, appear on the skin over the basal and intermediate segments of the digits. Additional complications affect the overall pattern and are often specific for different genera, but individual differences also occur, e.g. in the patterns on the digital pads where the original simple arches show various disturbances produced by elongation and deviation of the fasciculus principalis to one or

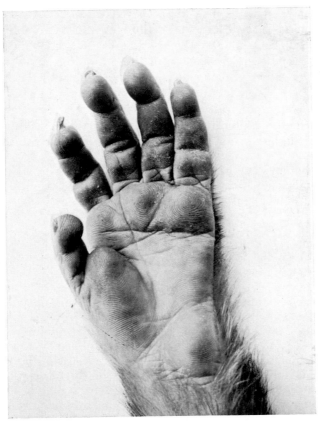

FIG. 25. Palmar surface of the hand of a Pig-tailed Macaque. (*Macaca nemestrina*) to show the flexure creases, palmar pads and dermatoglyphic configurations. $\times \frac{1}{1}$

other side, whereby looped patterns are formed. A further elaboration of the same process results in whorls.

Physiologically, the ridges appear in response to frictional stimulation and serve to prevent slipping between the palmar (or plantar) skin and the substrate. This is proved by the appearance of ridges even on the dorsal aspect of the terminal finger joints in the chimpanzee, where this area is applied to the substrate in the characteristic knuckle-walking way employed by this ape.

Ridge bearing skin, however, serves more than merely to assist grip. Being moist and abundantly supplied with sensory (tactile) nerve endings, it endows the extremities, especially the digits, with a high degree of sensibility, supplying the brain with information concerning the texture, size and other features of external objects, including food items. It also encourages the manual exploration of the environment and of the possessor's own body and those of its mates. Primates in possession of these faculties enter an entirely new and richer world of experiences. Coupled with the acquisition of stereoscopic vision a much implemented three dimensional view of the environment is attained. This intimate association between visual and tactile exploration has profound significance in the evolution of the higher primates (including man) (Elliot Smith, 1927)

"Under the guidance of vision the hands were able to acquire skill in action and incidentally to become instruments of an increasingly sensitive tactile discrimination, which again reacted upon the motor mechanisms and made possible the attainment of yet higher degrees of muscular skill. But this, in turn reached upon the control of ocular movements and prepared the way for the acquisition of stereoscopic vision and a fuller understanding of the world and the nature of the things and activities in it. For the cultivation of manual dexterity was effected by means of the development of certain cortical mechanisms; and the facility in the performance of skilled movements once acquired was not a monopoly of the hands but was at the service of all muscles. Skilful use of the hands was impossible without the appropriate posturing of the whole body. High co-ordination of hand movements and high co-ordination of the movements of the muscles of the whole body must go together. The sudden extension of the range of conjugate movements of the eyes and the attainment of more precise and effective convergence were results that accrued from this fuller cultivation of muscular skill." (Elliot Smith (1927) p. 152).

Transition from the brain structure of the tupaioids to that of frankly diurnal lemuroids was gradual. In its simplest expression it is seen in the brain of the small, primitive and still largely nocturnal mouse-lemurs (*Microcebus*) of Madagascar.

In *Microcebus*, the brain is the most primitive of any among living lemurs. The olfactory bulbs, though reduced in comparison with *Tupaia*, are better developed than in other lemurs and a fair proportion of pyriform cortex remains exposed laterally on the projecting tip of the temporal lobe, delimited by a distinct although shallow rhinal fissure. The neopallium is simply showing folding in two regions only. A deep sylvian fissure on the lateral surface and a triradiate (calcarine) sulcus on the medial surface of the occipital lobe. The posterior limb of the calcarine (retrocalcarine) is an axial folding of the visual (*striate*) cortex and highly characteristic of the Primate brain. Neopallial expansion is also implied by the elongation of corpus callosum and thickening of its rostral and caudal ends (genu and splenium) while concomitantly the fornix is reduced. As in *Tupaia* hippocampus and gyrus dentatus are still exposed strips of cortex extending down the medial aspect of the temporal lobe below the corpus callosum.

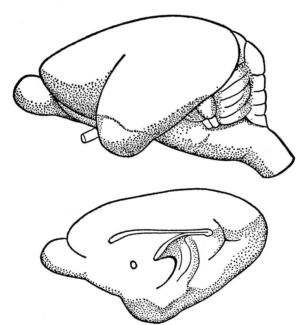

Fig. 26. Left side of the brain of *Microcebus murinus* (Lesser Mouse Lemur); and below; that of the right half of the medial aspect (cerebral hemisphere only). After Le Gros Clark (1931). ×3.

In the cortex are to be noted:

(a) Expansion of the parietal association area.
(b) Expansion of the temporal association area.
(c) Incipient expansion of area 8 (frontal).

All the above are "association areas". They provide the anatomic substratum for mental processes (association of ideas, memory etc.)

The thalamus, especially the lateral geniculate nucleus, shows also characteristic primate advances (see Le Gros Clark (1959) Fig. 113, p. 245).

In the larger, more typical and largely diurnal lemurs, the neopallial cortex is more richly folded and the olfactory areas further reduced. The basic pattern of folding consists of more or less longitudinal sulcation in the frontal and parietal regions, cutting across the cortical areas (e.g. s. rectus and s. intraparietalis). This contrasts with the higher primates, where there is a tendency for delineation of the histological areas by limiting sulci (e.g. s. centralis). A sulcus centralis, however, is present in the Potto's brain.

Cerebral features which indicate the necessity of aligning the lemurs and their allies with the monkeys are:

1. They agree in the possession of a true sylvian fissure – not present in other eutherians.
2. A sulcus centralis *may* be present (e.g. *Perodicticus*).
3. Motor cortex has similar structure and location.
4. Calcarine complex common to lemurs and monkeys.
5. Similarities in cerebellum.

Among the features in which the lemurine brain surpasses that of non-climbing mammals are the growth both of the cerebrum and the cerebellum The lateral lobes of the cerebellum are especially enlarged in adaptation to the necessity of maintaining side to side balance in the less stable environment of the trees compared with a ground living existence. Consequently, the vermis of the cerebellum becomes overshadowed by the lateral lobes. The cerebral hemispheres expand upwards, forwards, sideways and in the case of the temporal lobe, downwards and forwards. Backward growth of the occipital lobes carries them over the top of the cerebellum, which thus comes gradually to be completely hidden from above. Growth of the temporal lobe converts the sylvian fossa into a true sylvian fissure, so that the so-called insular or pararhinal cortex (i.e. the cortical cap of the striate body, one of the basal grey masses adjacent to the thalamus) is eventually buried. These cerebral improvements leave their mark on the bones of the brain case; the growth of the temporal lobe especially invading the region behind the orbit and in consequence of which the eyeball alters its position and no longer points to the side, but is directed forwards and pressed medially, compressing the nasal fossae – a feature especially well marked in *Tarsius*.

The brain of *Lemur* shows manifold advances on that of *Microcebus* both in gross morphology and histologically structural complexity. The olfactory parts are still somewhat in evidence and, though reduced in comparison with *Microcebus,* a small area is still visible laterally at the temporal pole, which otherwise has completely masked the pyriform area. The further expansion of the neopallium has necessitated folding of the cortex by a definite pattern of sulci, affecting primarily two regions only. A deep sylvian fissure or lateral sulcus is formed in the manner above mentioned. On the medial aspect of the occipital lobe a triradiate sulcus appears. This is the calcarine complex of which the posteriorly directed horizontal limb (retrocalcarine sulcus) is formed by a folding along the axis of the visual cortex and is highly characteristic of the primate brain.

Thus far *Lemur* resembles *Microcebus,* but additional sulci have now appeared, notably an oblique depression on the temporal lobe, parallel to the sylvian (parallel or superior temporal sulcus) and a further shorter (second temporal) below and behind this. Two longitudinal sulci have also appeared, one on the frontal (sulcus rectus) and one on the parietal area (intraparietal).

These new foldings show no relation to the histologically differentiated cortical areas, but tend to cut across them. The sulcus rectus traverses the middle of the frontal association area, and the intraparietal the parietal association area, extending thence to the fore part of the visual area.

In the larger lemurs, however, a small and irregular dimple sometimes appears near the dorsal end of the boundary between the motor and sensory areas. This appears to correspond to a well marked oblique sulcus (s. centralis) which delimits the two areas in monkeys (and also in carnivores). Occasionally this dimple also becomes a true sulcus centralis in some of the lorisoids where it is normal, e.g. in the potto (*Perodicticus*).

Another area of the brain of *Lemur* that shows advances resulting from improved visual capacity is in the lateral geniculate nucleus. Differentiation has resulted in distinct lamination of its constituent

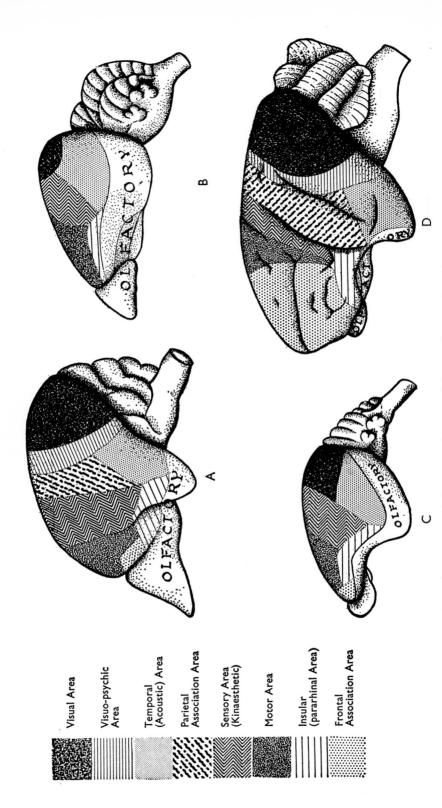

Visual Area

Visuo-psychic Area

Temporal (Acoustic) Area

Parietal Association Area

Sensory Area (Kinaesthetic)

Motor Area

Insular (pararhinal Area)

Frontal Association Area

Fig. 27. Diagram of the cortical areas of the cerebrum of A. *Tarsius*, compared with those of B. *Macroscelides*, C. *Tupaia* and D. *Lemur*. From Hill (1955)

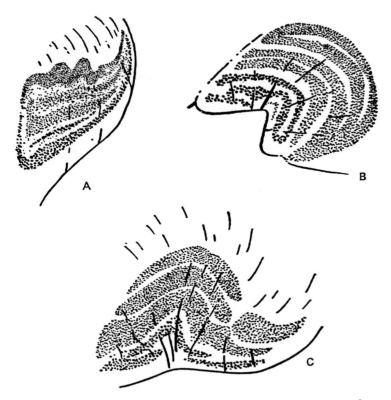

FIG. 28. Vertical sections through the lateral geniculate nucleus of A. *Lemur*; B. *Cercopithecus* and C. Man to show differences in arrangement of cell-laminae. From Le Gros Clark.

nerve cells, with the outer layers composed of large cells and the inner of small ones. The laminae moreover tend to become inverted at the dorsal and ventral margins, the fibres of the optic tract entering below and laterally, whilst from the concave mesial aspect the fibres of the optic radiation relay impulses to the visual cortex.

Further cerebral elaboration affects the brain of such lemuroids as the short-faced sifakas and indrises (family Indriidae) and also occurred in some now extinct (subfossil) monkey-like types such as *Archaeolemur*, a circumstance which suggests that the basic lemuroid stock, which became isolated in Madagascar, was well on the way to evolving independently by parallel evolution what, to all intents and purposes, was a true monkey as regards its ecological and biotopical status.

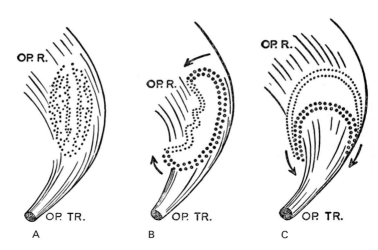

FIG. 29. Diagrams of transverse sections of the lateral geniculate nucleus and adjacent structures in an evolutionary series of Primates. A. primitive arrangement seen in *Tupaia*, where the cell layers are not sharply demarcated into laminae. B. arrangement met with in prosimians and *Tarsius*, where the laminae become inverted. C. showing the eversion of the laminae seen in all higher Primates. OP.TR. optic tract; OP.R. optic radiation. From Le Gros Clark.

6

The Tarsioid Phase of Cerebral Evolution

Although geologically speaking, there were tarsioids in existence at the dawn of the Tertiary Epoch, contemporary with the earliest lemuroids, they were strikingly better brained than the latter. This is evidenced by comparison of the skulls of those (e.g. *Necrolemur* and *Tetonius*) whose cranial remains are adequately represented for such a judgment. These relics indicate that their owners possessed visual and auditory equipment, as well as cerebral endowment, comparable to that of their sole surviving descendant – the small East Indian *Tarsius*, a rat-sized, long-legged, goggle-eyed sprite of largely nocturnal and arboreal habit that in its total anatomy, including its brain, displays a most remarkable combination of primitive and advanced characters. The advanced features place it on a higher plane than the prosimians and point in the direction of the monkeys. For example, *Tarsius* has lost the naked moist rhinarium, its nostrils are no longer slit-like (strepsirrhine), but rounded and haired up to their margins (haplorhine). Nasal chambers and facial vibrissae are reduced, the face being relatively flat and chiefly remarkable for the enormous, prominent and forwardly-directed eyes.

The brain (and hence the brain case) is rather globular in shape, for the hemispheres present a broadly simian appearance, having expanded sideways and also backwards and covering a greater amount of cerebellum than in the lemuroids. Olfactory bulbs are much reduced, being laterally compressed, while olfactory tract and pyriform cortex are correspondingly small, though the latter forms as much of the rounded temporal pole of the hemisphere as in the

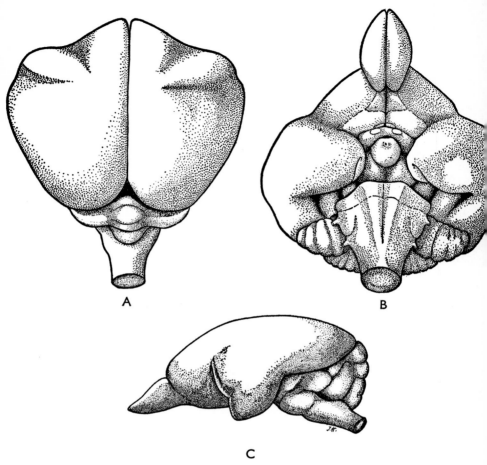

FIG. 30. *Tarsius syrichta*; upper, lower and left lateral aspects of the brain. From Hill (1955).

lemuroid brain. The expansive neopallial cortex surprisingly shows no trace of sulcation on its exposed surface, which is quite smooth and rounded, except for a short oblique depression at the site of the sylvian fossa anterior to the rather bloated temporal pole. On the mesial aspect, however, a deep triradiate calcarine fissure is present, located in a depression that accommodates the large optic lobes (superior corpora quadrigemina) of the mid-brain. Whilst the hippocampal formation (hippocampus and dentate gyrus) agrees with that of monkeys in being exposed only at its lower extremity; the corpus callosum is primitive, resembling that of a lipotyphlan insectivore in its shortness and lack of fore and aft thickening. The anterior commissure is, on the other hand, relatively large.

Histological differentiation of the neopallial cortex (Woollard, 1942), shows advances on the lemuroid level. Notable is the enormous extent of the visuo-sensory cortex (area 17) and its functionally associated peripheral zone (area 18 of Brodmann) which together account for almost half of the total surface of each hemisphere. To this expansion of the visual areas the growth of the occipital pole over the cerebellum is due. This is accompanied by an extension into the occipital pole of the hollow interior of the hemisphere (lateral ventricle) to provide an occipital horn – a feature characteristic of the simian brain, but which is lacking in that of lemurs. Other cortical areas do not show much advance beyond the condition in *Microcebus*.

The thalamus also recalls that of *Microcebus* in size and proportions but its by-product, the lateral geniculate body, is much larger proportionately, as is also the case with the adjacent nucleus within the thalamus which sends its fibres to area 18 of the cortex. This nucleus is the precursor of the much larger structure seen in simian brains as the pulvinar. Nerve cells of the lateral geniculate nucleus are arranged in distinct laminae, which are inverted as in *Lemur* (Le Gros Clark, 1930, 1932).

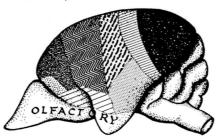

FIG. 31. *Tarsius bancanus borneanus*; left lateral aspect of the brain to indicate Brodmann's areas. For key see Fig. 27.

The cerebellum of *Tarsius* is quite primitive, the lateral lobes being small and the vermis relatively large, representing a stage even less advanced than in the brain of *Microcebus*.

This curious combination of archaic and advanced characters of the tarsioid brain poses many problems of interpretation, especially in view of the contemporaneity of its phylogenetic development with that of the Lemuroidea. Basically the brain is truly primitive, but has had grafted upon it a vast structural complication arising from a massive elaboration of the visual functions. What use this can be to nocturnal creatures provided with a retina which, though having a macula, is constructed for night vision and not for sharp, critical discrimination is somewhat of a mystery insofar as there is no corresponding improvement in the motor and other non-visual areas of the cortex. But these are problems beyond the scope of this work. However, taken in conjunction with anatomical advances elsewhere in the body, it is clear that *Tarsius*, on the whole, portrays a phylogenetic horizon intermediate between the prosimian and simian levels of primate evolutionary progress.

7

The Simian or Pithecoid Phase

A significant behavioural development that was largely responsible for further cerebral advance was, in monkeys, the emancipation of the pectoral limbs from bearing body weight, at least at certain periods. This has freed them and particularly the sensitive hands, for exploring the environment, when more precise information on shapes, textures and so on could be passed to the sensory areas of the brain. On the motor side greater mobility of the hand and forearm followed and particularly more independent movement of individual digits. Examination of and discovering the nature of its own body and that of its neighbours also resulted in a flood of new stimuli to be transmitted to the cortex. Incidentally, this tactile exploration of the fur of other members of its own species had a significant bearing on the development of social behaviour and social cohesion, through the phenomenon of social grooming.

These changes have occurred *pari passu* with changes in body posture and concomitant skeletal adjustments. The adoption of an erect or suberect body stance, which occurs when a primate squats on its haunches, necessitates changes in the poise of the head so that the eyes may still be directed forwards and this direction maintained also during quadrupedal progression. To bring this about, the articulation between skull and spine is brought downwards and forwards whence the position of the foramen magnum is also altered accordingly. At the same time, the shortened muzzle is brought downwards beneath the orbits and forepart of the brain case resulting in the appearance of a forehead, maybe low and backwardly inclined, but nevertheless distinct and varying in ex-

c

tent according to the degree of development of the frontal lobes of the cerebral hemispheres.

The lowliest degree of simian cerebral advance is seen in the marmosets (*Hapale* or *Callithrix*) and their relatives of the family Hapalidae or Callithricidae. In body size these diminutive monkeys are comparable to *Tarsius* or the smaller lemurs, yet the brain weight is thrice the amount. The increase affects mainly the cerebral hemispheres which are voluminous, with large frontal lobes and occipital lobes that extend backwards so as completely to cover the cerebellum and medulla from above. A posterior horn of the lateral ventricle extends into the occipital lobe. Olfactory bulbs are reduced further than in *Tarsius* and the same applies to other parts involved in olfactory reception (olfactory tubercle and pyriform cortex, the latter being now almost entirely relegated to the basal and mesial aspects of the hemisphere). A short, shallow but distinct rhinal fissure, however, persists. The anterior and fornix commissures are small, contrasting with an enlarged corpus callosum, which shows anterior (genu) and posterior (splenium) thickenings in correlation with the voluminous neopallium. Except for a deep sylvian fissure and a retrocalcarine sulcus that extends back almost to the occipital pole, deeply incising the visual (or striate) cortex, the surface of the neopallium is smooth. Some specimens, however, exhibit a faint representative of the parallel sulcus on the bulky temporal lobe.

Histological differentiation of the hapalid neopallial cortex is much in advance of that found in *Lemur* or *Tarsius*. There is a considerable expansion of association areas, notably in the frontal and parietal regions, as well as further cellular differentiation in

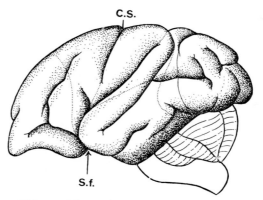

C.S.

S.f.

FIG. 32. Brain of an Old World monkey (Purple-faced Langur, *Kasi senex senex*) from the left side. C.S. central sulcus; S.f. Sylvian fissure.

their intrinsic structure. Visual cortex is now thrust farther back with the enlargement of the parietal area and in this too lamination has proceeded further, though not to the degree met with in some of the larger New World monkeys, where the lamination is further advanced than in the great apes and man.

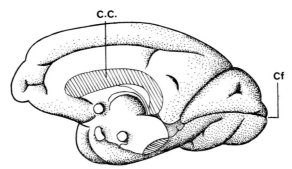

Fig. 33. Right cerebral hemisphere of an Old World monkey (Purple-faced Langur, *Kasi senex senex*), viewed from the inner aspect. CC. corpus callosum (in section), Cf. calcarine fissure.

The cerebellum of the marmoset exhibits considerable lateral expansion of its lateral lobes compared to that of *Lemur*, but the floccular lobes are relatively reduced.

In the remaining New World monkeys (family Cebidae) and in all those of the Old World (families Cercopithecidae and Colobidae) the hemispheres are further enlarged and their surface more richly convoluted. On the exposed surface the large and deep sylvian fissure is accompanied by the now deeper parallel sulcus on the temporal lobe. Dorsally these two appear to unite and to be continued almost to the dorsal edge of the hemisphere. A well marked central sulcus incises the upper margin some distance anterior to this conjoined fissure and is continued obliquely downwards and forwards until it almost meets the sylvian. It thus demarcates the frontal from the parietal lobe and serves to separate the motor from the general sensory areas of cortex. The frontal lobe is marked by two sulci, a horizontal sulcus rectus and a triradiate arrangement between the posterior end of the latter and the central sulcus, usually known as the sulcus arcuatus. The parietal lobe is folded to produce an intraparietal sulcus, a basically oblique depression whose upper end cuts the dorsal border of the hemisphere to carry it on to the mesial aspect. The incisure is at or near the site where the upper end of the parallel sulcus terminates. The occipital lobe is variously marked, but the most important feature is a vertical deep limiting sulcus in front of the visual cortex. This is the sulcus lunatus which, in view of its universal presence in these monkeys, is often known as the simian sulcus. Externally, the occipital lobe of most simian hemispheres, though extensive, is characteristically quite smooth between the simian sulcus and the occipital pole. An axial folding of the visual cortex on the mesial aspect produces a deep calcarine sulcus; it is joined at its anterior end by the mesial continuation of the intraparietal sulcus. A longitudinal sulcus (often branched) marks the mesial aspect halfway between the upper border and the corpus callosum; it is named calloso-marginal sulcus.

The thalamus in monkeys is considerably more differentiated into distinct nuclear masses than in the prosimian or tarsioid brain. Particularly characteristic is the expansion of its postero-dorsal nucleus which now appears as a surface projection or rounded tubercle called the pulvinar. In the adjacent lateral geniculate body the lamination of the nerve cell layers has become more distinct and the edges of the laminae have become everted (instead of inverted as they are in prosimians and *Tarsius*). In the process of eversion a ventrally directed concavity or hilum is produced wherein are received the fibres from the optic tract. Projection fibres to the visual cortex (optic radiation) depart towards the cortex from the convex dorsal aspect of the nucleus.

The cerebellum, especially its lateral lobes, enlarges, particularly in the larger monkeys and apes, in proportion to the growth of the neopallium. Lateral lobes share a two-way connection with the motor areas of the cortex, for functional co-operation with the cortex plays an important role in control of the voluntary muscles. Further differentiation of the lateral lobes involves complicated folding and fissuration, but also development of secondary lobulation. At the same time the floccular lobe retains its distinctness – a heritage from lower mammals – and is lodged in the subarcuate fossa, a deep excavation in the petrous bone.

8

The Hominoid Phase

The most advanced primates are those comprising the superfamily
Hominoidea, a group composed of three families: Hylobatidae
(gibbons), Pongidae (the great apes) and Hominidae (man). For the
present purpose all these may be taken together for, in their cerebral
endowment, they show progressive development based on the
changes already laid down in the simian brain.

Most striking is the further enlargement of the cerebral hemisphere accompanied by increased folding of the cortex. In some
places the folding has involved secondary replication of areas already marked by fissures. Hereby certain convolutions are hidden
in the depths of some of the major sulci such as the sulcus centralis
and sulcus calcarinus. An analogous process is seen in relation to
the insular cortex which, though remaining partially exposed in the
simian brain, becomes progressively masked by the overgrowth of
lobe-like flaps or opercula derived from the expansions of the
parts of the frontal, parietal and temporal lobes where they approach
over the stem of the Sylvian fissure. Only in the human brain is the
operculation complete, though closure is virtually attained in the
brains of the great apes.

Total increase in brain size is, of course, related in part to increased body weight in the great apes and man, but this is not the
whole story. Cerebral growth is strikingly evident in the frontal
lobes, a feature which appears to have resulted in a downward
bending or kyphosis of the brain axis (and in consequence that of
the base of the skull also) as pointed out by Hofer (1969a). Fissuration
of the parietal lobe has made marked advances even in the gibbon's
brain and is further developed in the brains of the great apes and

man, where expansion of the parietal association area has thrust the visual cortex farther back and to some extent round the occipital pole on to the mesial aspect. Displacement of the visual cortex results in the almost complete effacement of the simian sulcus, except for an inconspicuous depression not always to be identified in the welter of new secondary and tertiary foldings in this region. An intermediate condition between the degree of parietal expansion seen in the ape brain and that characteristic of man was evident, as judged by endocranial casts, in the brains of the Pleistocene fossil ape-men of the subfamily Australopithecinae.

Histological differentiation of the cortex has also been apparent in the progress from monkey to ape and ape to man. Especially notable is the increase in total number of nerve cells to unit volume of grey matter which is 50% higher in the visual cortex of man than in the chimpanzee. This is readily inferred as conferring a great degree of association connections between the different control areas. Curiously enough, the lateral geniculate body is rather more primitive in the brain of the great apes and man than in the gibbon or in monkeys, for the laminae show incomplete eversion. This may mean that the latter are independently specialized in this direction.

The cerebellum of the Hominoidea exhibits a further advance in the fissuration and subdivision already referred to in the simian brain. Floccular lobes have, however, undergone further reduction and are no longer housed in a subarcuate fossa which is indeed absent in the petrous bone of the great apes, though still present in the smaller gibbons.

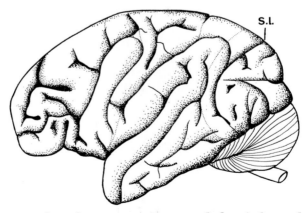

Fig. 34. Brain of a chimpanzee (*Pan troglodytes*) from the left side. S.l. sulcus lunatus.

9

Modes of Progression
in Primates

Primitive mammals were quadrupedal and terrestrial, a state of affairs inherited from their reptilian ancestors. Their higher rate of metabolism, however, which involved the capability of more rapid locomotion, led to a lengthening of the limbs and their endowment with sufficient power to raise the body off the ground during the whole of their waking life – in contrast to the reptilian attitude of resting the belly upon the substrate with the limbs sprawled out laterally, except during the sporadic spurts of activity. In this primitive quadrupedalism all four limbs are approximately of equal length and the body is carried horizontally.

With the adoption of an arboreal mode of life, various changes in the mode of progression are feasible. Several of these have been exploited by members of the primate order, with correlated structural changes in the limb skeleton and associated muscles. Even at the level of the menotyphlans some successful experimentation has taken place. Many of the tree-shrews are still mainly terrestrial or spend much of their time on the ground or in low scrub. Others, like *Ptilocercus* and *Tupaia minor* are more exclusively arboreal. Though their limbs have changed little as regards proportions, the manus and pes show changes in the greater independence of the digits, the improvements in their tactile sensibility and in a greater degree of separation of the first digit, both on hand and foot – all adaptive features of use in progression in an arboreal environment.

A totally different means of progression has been developed in the elephant shrews (Macroscelididae) in spite of their retaining a terrestrial habitat – a method which has been exploited equally successfully by a number of non-related mammalian types, both in terrestrial (jerboas, jumping mice, kangaroos) and arboreal (tree kangaroos) environments. This type of progression is known as saltation; it involves the elongation of the hind limbs, more especially of certain segments of the limbs, combined with hypertrophy of the muscles, the extensor muscles being specially developed. These changes impart to the limb the power of rapid and powerful extension which propels the animal forwards in a leap or bound. Forelimbs remain short or may even be reduced, so that the bodily stance tends towards a semi-erect posture. The tail is generally strengthened and may be used as a prop, so that when at rest the body is supported on a tripod. This modification has also been exploited by several primate lineages, both fossil and recent.

Evidence from comparative anatomy and paleontology supports the view that even in the earliest precursors of the primates there was developed some degree of grasping or clinging power in the extremities, in addition to their basic locomotor functions. This is already manifest in the most primitive of the marsupials (opossums) as well as some more advanced ones (phalangers) and is incipient, as we have seen, in tree-shrews. In addition to the local modifications in the hand (curved metacarpals and phalanges, separation of pollex etc.) there are concomitant changes affecting other parts of the limb adapting it to the grasping function. Free movement in all directions at the shoulder joint is an important *sine qua non* associated with a well developed clavicle that ensures the shoulder from being hampered in its movements by proximity to the chest wall. Complete separation of the two bones of the forearm (radius and ulna) permitting rotation from the prone to the supine position permits freer uses of the hand in different positions both in grasping objects in the environment, exploration and manipulation of food items. In climbing, the hands can then be used to aid the body in progression by hauling it forwards or upwards – a process which is assisted by certain developments of the musculature of the pectoral girdle (Ashton and Oxnard, 1963, 1964a, b).

The hind limb functions somewhat differently and this is reflected in its anatomical details. Whereas the pectoral girdle is adapted for free range of movement, the pelvic girdle is designed for stability in weight transmission, being soldered to the spine and by early consolidation into a single bone (*os innominatum*) of the three structural elements of which it is composed. This, combined with stability at the hipjoint, ensures the required rigidity for those occasions when, with the hands emancipated, the whole of the body weight must be transmitted through the hind limbs. For the same reason mobility between the shank bones (tibia and fibula) is greatly limited compared with their counterparts in the forearm. Nevertheless, the distal segment of the limb (the pes) retains its primitive pentadactyl character with freely separated digits or widely divergent hallux and similar cutaneous modifications as in the hand.

The most primitive primates such as *Microcebus* have not advanced beyond the tree-shrews in their mode of arboreal progression – a simple quadrupedal running, jumping and climbing. Most other primates have specialized to some extent in their methods of progression, all the possible modes of translation having been exploited by one or other lineage, in some cases more than once, leading to some earlier errors in the estimation of affinities.

Primates may be classified on locomotor habits, following Napier and Napier (1967) into the following primary categories:

1. Quadrupedal walkers, runners and climbers.
2. Vertical clingers and leapers.
3. Brachiators.
4. Bipedal walkers and cruriators.

Quadrupedalism

This is susceptible to several modifications and specializations so that Napier and Napier subdivide this category into the undermentioned. (Their order is changed here to a more logical evolutionary sequence.)

(i) Branch running and walking type
(ii) Slow climbing type
(iii) Ground running and walking type (cursorial)
(iv) Semibrachiators (New World type)
(v) Semibrachiators (Old World type)

The principal component of the quadrupedal type of locomotion, whether in the trees or on the ground is a gait involving alternate use of each of the four limbs, with the pectoral and pelvic pair more or less equal in length, so that the body is carried horizontally. In the trees the hands and feet provide stability by virtue of their prehensile function. Propulsive acts such as springing, jumping and leaping are products of this mechanical set up. Quadrupedalism also accommodated vertical progression by the hand over hand heaving upwards of the body to a higher situation (climbing) assisted by subsequent follow up of the hind limbs. A large number of primate genera fall into this category so far as locomotor functions are concerned, notably the marmosets, capuchins (*Cebus*), *Cercopithecus* and some lemurs as well as *Tupaia*.

It is an easy step from these arboreal adaptations for a primate to return secondarily to a terrestial habitat, either part-time or completely. This has recurred in several evolutionary lines either as a pioneering effort in exploiting new food sources or more likely from pressure of circumstances in areas where deforestation has occurred through climatic changes. The phenomenon is best exemplified by the savannah dwelling baboons of the genus *Papio*, certain guenons of the *Cercopithecus aethiops* group and the Hussar monkeys (*Erythrocebus*). Some cliff and rock dwelling forms also fall into this category, notably most macaques (*Macaca*) and the geladas (*Theropithecus*). The position of the drills and mandrills (*Mandrillus*) is remarkable insofar as they are forest dwelling forms that have embarked upon a terrestrial existence on the forest floor, thus occupying an otherwise vacant ecological niche as far as primates are concerned. They are nevertheless good climbers and still resort to the trees for shelter on alarm and for sleeping (Hill, 1970).

In terrestrial running and walking, the grasping functions of the extremities are not involved. There is a tendency to shortening and increased robusticity of the hands and feet with shortening of the digits – well exemplified is *Erythrocebus*. A digitigrade posture is adopted, with the body weight transmitted through the distal half of the palms and soles, the proximal half (heel) being raised and held in line with the forearm and wrist joint in extension (Jones, 1967).

A very specialized modification of arboreal quadrupedalism is that exhibited by the nocturnal lorises and their allies (*Loris, Nycticebus, Perodicticus* and *Arctocebus*). This consists of slow hand over hand and foot over foot progression – usually with only one extremity freely groping forwards while the other three maintain a rigid grasp of the neighbouring branches. The phenomenon is reminiscent of that employed by the South American sloths (Bradypodidae) and, like them, the lorisoids are able to progress in a suspended position on occasion. There are also parallel structural developments, especially in the specialization of the blood vessels of the limbs (retia mirabilia) and preponderance of red (slow prolonged action) muscle fibres in the flexor muscles of the limbs (e.g. Wislocki and Straus (1932)). The hands and feet, however, though modified are still recognizably primate in structure and appearance, differing thus markedly from the same members in the sloths. Modifications are in the direction of enhanced grasping power; by the acquisition of a forceps like function (e.g. as in chameleons) with the two limbs of the forceps formed by digits I and IV, the others tending to be reduced, especially digit II, which in *Perodicticus* is represented by a mere tubercle. Hypertrophy and free abduction of the first digit permits a 180° angle to be formed with the axis of digit IV. The tail is much reduced or lost.

In spite of the normally tardy movements in these lorisoids, the animals are still capable of fairly rapid translation when pressed, but they are quite incapable of leaping.

Semibrachiation

This is a pattern of progression wherein the forelimbs take a major share in propulsion without being exclusively involved. In the form adopted by New World monkeys such as spider monkeys (*Ateles* and *Brachyteles*) woolly monkeys (*Lagothrix*) and howlers (*Alouatta*) the fore-limbs are extended above the head and the body suspended by them with assistance from a prehensile tail. The body is swung forwards with some aid from the hind limbs. Quadrupedal walking or running is alternatively used for much of the locomotor requirements, but leaping is largely in abeyance.

Old World exponents of semibrachiation include the six or more genera of the family Colobidae (the Asiatic leaf monkeys and the African *Colobus*). Hand over hand progression is seldom used, but the fore-limbs are raised above the head and forwards to grasp a branch or check momentum. Leaping is more in evidence than in the New World monkeys. Some structural modifications of the fore-limb occurs – elongation of the palm and fingers and reduction of the thumb (reduced to a tubercle in *Colobus* as in the New World *Ateles* and *Brachyteles*). The hand thus is adapted to function as a hook rather than a grasping tool.

Vertical Clinging and Leaping

This important locomotor category is, in many ways, the reverse of the slow quadrupedal progressions of the lorises. The contrast is well shown in the galagos, relatives of the lorises that have diverged in their mode of arboreal progression which is by rapid saltation or leaping from one fixed point to another. The main impetus for movement resides in the hind limbs which are greatly lengthened and provided with powerful hip, thigh and calf muscles, whereas the fore-limbs are short and take little part in locomotion. The tail is long and used as a balancing organ.

In addition to the galagos, vertical clinging and leaping is exploited by several Malagasy lemuroids (*Hapalemur*, *Lepilemur* and the specialized indrises of the genera, *Indri*, *Propithecus* and *Avahi*) and also by *Tarsius*.

In all of these the body is held vertically when at rest, pressed to the main trunk of a tree or to a vertical branch. Progression involves a spring, leap or jump from one vertical branch to another and may be quite rapid when a succession of such leaps takes place. Hands are used solely for grasping and take no part in propulsion. The tail, when present, may be used as a prop (e.g. in *Tarsius*) or for balancing. In *Indri* it is reduced to a stump.

If placed on the ground, primates in this locomotor category adopt a partial or completely erect, vertical attitude or stance. They progress by bipedal hopping when moving rapidly, but, like kangaroos, assume a quadrupedal gait when progressing slowly. Leaps from the standing position may be of relatively immense distance compared with the size of the animal (see Hall-Craggs (1965a, b)).

Brachiation

This is a specialized mode of locomotion seen in greatest perfection in gibbons (family Hylobatidae). Gibbons are long armed, tailless primates capable of very rapid arboreal locomotion by swinging pendulum-wise suspended by the arms. Impetus to the pendulum (i.e. the body) is supplied by the hind limbs, which are pressed against a support to initiate the swing. By release of the hand-hold, alternately left and right, the released hand moving forwards to a new support, translation through the forest can be very rapid. On the ground an erect attitude is adopted, supported on the extended hind limbs. Progression is then by a balanced bipedal gait, with arms held aloft. Bipedal walking along branches is also employed by gibbons, hands being free to explore the environment, collect food or simply to steady the movements.

In the large apes (orang, chimpanzee, gorilla) a modified brachiation is adopted when in the trees. The orang is the best exponent, being the most arboreal. Gorillas and chimpanzees are more terrestrial and show further modifications for this type of existence.

In this modified brachiation the extended forelimbs still play the major part in suspending the body and propelling it through space. Hind limbs provide some support below and may play some part in heaving forwards the heavy torso. On the ground, a peculiar pattern of support and locomotion is evolved and referred to as "knuckle-walking". This is a modified quadrupedal gait where the long forelimbs are used somewhat in the manner of crutches. Combined with the relatively short legs this results in the body being held in an oblique posture midway between the horizontal and the vertical. This secondary return to the ground seems to have become needful with increased body bulk, which hampers agile arboreal activity. The return to ground living, however, did not effect a reversal of the evolutionary adaptations of the extremities. The hands, particularly, are no longer capable of supporting the weight through the flat palm. Instead, the digits are flexed and body weight, both in standing and walking, is transmitted through their dorsal aspects, i.e. walking is by means of the knuckles which, as a result of the changed circumstances, develop callosites on the areas of contact with the substrate. These changes are less extreme in the foot, but here too the toes are usually flexed to some degree. Inconvenience is experienced in the use on the ground of a foot preadapted for arboreal progression. Such a foot has the sole somewhat inverted in order to grasp the side of a branch. This is especially well seen in the orang. For perfect adaptation on the ground, the foot must be everted to enable the sole to make free contact, otherwise the body weight is transmitted through the outer edge of the foot. Such eversion is maintained to a better degree in the chimpanzee and best of all in the gorilla, the least arboreal of the great apes. These are therefore better adapted than the orang for a return to semiterrestrial life. The change involved modifications in bones, joints, ligaments and musculature.

Bipedalism (Cruriation)

Many monkeys on occasion adopt an erect attitude supported solely on their hind limbs. This is usually quite momentary, as in scanning over long grass or to gain a better view in the event of the possible presence of predators or other adverse elements. Still less frequently is bipedal locomotion adopted, though this does occur, e.g. when both hands are occupied in carrying off food pillaged from native cultivation or where a female supporting an infant with one hand has the other occupied with snatched foot items. These instances do, at least, indicate the potential for the development of true bipedalism. Further developments in this direction are shown in the chimpanzee and gorilla, which are both capable of more prolonged bipedal progression, although their normal terrestrial travel is quadrupedal. In the gibbons, however, as we have seen, more perfection in bipedal progression is seen insofar as this is their normal method of transit on the admittedly rare occasions when they find themselves on the ground.

In true bipedalism, which apart from the above mentioned exceptions is only found in man, the body is habitually held perpendicular to the substrate. The hind limbs are extended at the hip and knee joints so as to transmit the whole of the body weight. The hind limbs move alternately to produce a gait characterized by striding. Propulsion involves a heel and toe interchange. There are many structural modifications in the foot. Besides the perfecting of eversion, a re-arrangement of the pedal skeleton has evolved the formation of plantar arches, whereby both longitudinal and transverse concavities are formed in the centre of the sole. As a result the supporting pillars of the longitudinal arch are formed respectively by the heel bone (calcaneus) and the heads of the metatarsals (which are level with the site of the distal or interdigital touch pads). Modifications of the musculature ensure the maintenance of the arches. Another unique development is the shift of the first digit (hallux) into alignment with the other toes whereby the power of grasp has been lost. At the same time, the lateral toes are tending to atrophy, especially digit V which often shows reduction in the number of its bony elements and even the loss of its nail.

10

Comparative Cranial Anatomy of Primates

The evolutionary changes that have occurred in the head skeleton of the primates reflect in large measure the changes which have been recorded in the brain and organs of special sense. These changes are represented today in the differences observed in the skulls of a representative series of modern primates from tupaioids to man.

They may be summed up in the following terms:

(a) the tendency towards a progressive enlargement of the brain case.

(b) a parallel tendency to a reduction or at least shortening of the facial skeleton in correlation with lessened importance of the olfactory function.

(c) enlargement and shift of the orbits in connection with the increased importance of vision.

(d) alteration of the position and angle of articulation of the skull with the vertebral column in association with the adoption, in varying degrees, of an upright (or orthograde) posture in an arboreal environment which has freed the forelimbs for prehension and grasping – a function formerly carried out by the teeth and lips.

Enlargement of the Brain Case (Neurocranium)

The brain case of the Insectivora is rather narrow and elongated; it lies directly posterior to the facial skeleton and the axes of the two parts are more or less in line. This situation is still maintained in the monotyphlan insectivores. With the enlargement of the brain that is seen in the lemurs, even the most primitive forms like *Microcebus,* the brain case assumes a more globular form as the brain expands in all directions except downwards. This expansion is usually accompanied by some down-bending (kyphosis) of the facial axis whereby an angle is formed, open downwards, with the basicranial axis. Another incidental effect of the neuro-cranial expansion is to provide a broader attachment, in the temporal region, for the muscles of mastication. This renders unnecessary, in many cases, the development of additional bony crests for the attachment of these muscles, though this depends to a large extent on the dietetic habits of the species concerned.

Reduction of the Facial Skeleton (Splanchnocranium) and Orbital Changes

The facial skeleton is basically formed by the bony framework of the nasal fossae and the tooth bearing bones (maxilla and mandible). As long as olfaction held primary sway the nasal fossae were located immediately anterior to the brain and were roomy and complicated internally by elaborate scroll-like bones upon which the lining was stretched to provide a maximum sampling service – the typical mammalian pattern. The fossae are separated by the bony palate from the mouth cavity, so that this too lies largely anterior to the neurocranium. Moreover, in insectivores especially, the muzzle is elongated and the long jaws give ample room for a series of formidable teeth adapted especially for the capture and crushing of living insects, including those with relatively hard chitinous exoskeletons.

With the increasing importance of vision in the lower primates, the lessening dependence on the sense of smell permitted the shrinking of the nasal fossae, which show a tendency both to antero-posterior shortening and transverse compression. This allowed the enlarging orbits to move forward from the temporal region and to be directed in an increasingly frontal axis (see above, p. 35). As the orbits enlarge, they become surrounded by a complete bony ring by the development of an articulation between processes from the frontal bone and the zygomatic. Isolation of the orbits from the temporal fossae (occupied by the temporal muscle) is somewhat further advanced in *Tarsius,* where a flange of bone grows inwards from the orbital ring. In simians this flange is enlarged so as to shut off the orbit completely except for a narrow slit – the inferior orbital fissure.

The zygomatic arch itself remains relatively slender but, unlike

that of many insectivores, it forms a continuous bony bar between
the temporal bone and the maxilla, serving to give attachment to
one of the muscles of mastication (the masseter).

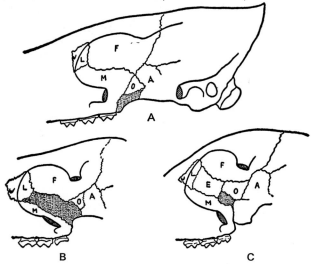

FIG. 35. Diagram to show the bony elements contributing to the orbito-
temporal region; stippling indicates the extent of the palatine bone.
A. Primitive mammalian condition with wide fronto-maxillary contact.
B. Condition occurring in the Tupaioidea and recent Lemuroidea where
   the orbital plate of the palatine extends forwards to meet the lachrymal.
C. Condition in Lorisoidea, *Tarsius* and the Anthropoidea where the
   ethmoid separates frontal from maxilla and the palatine from the
   lachrymal.
A. alisphenoid: E. ethmoid; F. frontal; L. lachrymal; M. maxilla; O. orbito-
sphenoid.

Changes in Cranio-Vertebral Articulation: Effects of Orthograde Posture

In quadrupedal mammals of average type, e.g. the dog, the foramen
magnum, through which the brain stem passes into the spinal cord,
faces almost directly backwards. Flanking the foramen are the two
occipital condyles; these are articular surfaces which are opposed
to corresponding cavities on the first cervical vertebra (atlas) in the
atlanto–occipital joint. The vertebral column being carried horizontally, the head is therefore held in alignment therewith.

In primates the direction of the foramen magnum and hence also
of the atlanto–occipital articulation has shifted downwards to a
varying degree. This displacement is usually regarded as an adaptation to the more vertical position of the trunk, the head alignment

being thereby maintained in a horizontal plane as most suitable for the effective use of the eyes.

The vertical posture of the trunk is itself, at least partly, the result of enlargement of the cerebrum, especially in the occipital region; the consequent expansion of the occipital region of the skull determining the downward shift of the foramen magnum. The nuchal region of the skull – that part of the occiput which receives the attachment of the muscles which haft the skull to the vertebral column immediately dorsal to the foramen magnum – likewise changes its direction progressively from posterior to inferior. Even in *Tupaia* a vertical position of the trunk is occasionally adopted. Wood Jones (1916) has argued that uprightness of the trunk is an ancient feature.

In many prosimians (e.g. Lorisidae, Lemuridae, Indriidae and in *Tarsius* and the fossil *Necrolemur*) a vertical or sub-vertical resting posture is habitual, whence their being grouped in the locomotor category of "vertical clingers and leapers" (See previous Chapter). All these exhibit a globular brain case with projecting occiput and the foramen magnum placed well forwards on the cranial base.

In all simian and hominoid genera a vertical body posture is maintained, for even where locomotion remains quadrupedal, as in for instance *Macaca* and *Cercopithecus,* long periods of resting in a sitting posture, especially during nocturnal sleep, are assumed. In New World monkeys (e.g. *Cebus, Saimiri*) sitting involves squatting upon the flexed hind limbs, the body weight being carried through the soles. In sleeping, New World monkeys lie down on their sides. In Old World forms a further specialization is found; the body weight, in sitting, is transmitted to the substrate through the ischial bones of the pelvis. An adaptation of the ischia, which exhibit varying degrees of expansion, occurs in the suprajacent skin which is modified by increased cornification and loss of hair to produce bare sitting pads (ischial callosities) (Washburn, 1957). Callosities vary in size, being small in *Cercopithecus* and gibbons and large in macaques and baboons, where in males the two are confluent and surrounded by a variable amount of hairless but uncornified skin. These differences, however, are indications of taxonomic affinity rather than correlated with habit or ecology; nevertheless, a general trend for larger callosities in ground or rock-dwelling species as compared with arboreal types seems to be manifest.

The balancing of the skull upon a vertically aligned vertebral column involves other factors than brain expansion. The size of the jaws and bulk of the muscles of mastication have also to be taken into account. Possession of powerful teeth subsumes a certain forward projection of the facial structure in which they are implanted (prognathism) and also bulky muscles of mastication to assist the jaws in their work. This has to be balanced against the backward expansion of the occipital region of the neurocranium. This situation prevails in *Lemur* and in consequence the overbalancing of the prognathous face is overcome by the existence of powerful nuchal muscles. In the short-faced indrises, on the other hand, the balance is effected by the greater forward situation of the condyles, a state of affairs even better exhibited by *Tarsius*, where the foramen magnum faces directly downwards and the condyles are located at approximately the level of the junction of the posterior and middle thirds of the cranial base. Secondary enlargement of the masticatory apparatus (teeth, jaws, muscles) from adaptation to a highly vegetarian diet, for instance in *Gorilla*, evokes the development of a powerful set of nuchal muscles to offset the imbalance due to the prognathism and the fact that the occipito–atlantal pivot is located farther back than in man. Such imbalance is of an even greater degree in the heavily jawed baboons; in these, as in *Gorilla*, there is an enormous expansion of the nuchal area for the accommodation of the dorsal neck muscles. The expansion is such that the cranial surface above is insufficient for the purpose, so that a bony flange (nuchal or lambdoid crest) is thrown out to increase the area for their attachment.

Nuchal crests are functionally analagous to the median sagittal crests often found in the dome of the neurocranium, but the two structures are independently acquired.

Sagittal crests depend for their existence on the size of the temporalis muscles. Where the jaws are heavy and powerful the temporalis muscle expands its attachment to the crown of the skull, where it meets its fellow, forming a bulge on each side of the middle of the crown with a groove between, e.g. in the Crab-eating macaque (*Macaca irus*). In extreme cases even this accommodation is inadequate, in which event a flange is thrown up in the midline, whereon the two temporal muscles can extend their attachment, as in *Gorilla* (Angst, 1970). Nuchal crests and a sagittal crest occur together in *Gorilla* and become confluent at the lambda; but the two flanges owe their formation to independent mechanical forces and so may occur separately. A sagittal crest may develop independently of a nuchal crest, e.g. *Nycticebus* (Seth, 1964) and *Colobus verus* (Vogel, 1962). Conversely a nuchal crest may exist in the absence of a sagittal crest, e.g. in *Colobus polykomos* (Vogel, 1962).

Another cranial modification which appears to go hand in hand with the displacement of the foramen magnum is the development and progressive increase in the kyphosis of the cranio–facial axis – already briefly alluded to (p. 70). In primitive insectivores, the face comprises approximately half of the total length of the skull which thus, for example in *Tenrec*, takes on what Frechkop (1949) terms a tubiform morphology. In this condition, the basicranial and basifacial axes are in line with each other, i.e. the angle of kyphosis is 180°. This must be an acquired condition since it involves a straightening out of the kyphosis which is present during early embryonic life in all mammals. To some extent, therefore, the down bending of the facial axis on the basicranial axis, so characteristic of all primates (though not confined to them), may be regarded as the retention of an embryonic condition. On the contary, it may be interpreted as a new evolutionary trend induced by biomechanical forces involved in adapting the skull to the needs of the orthograde posture, rebalancing of the cranium on the spine and the displacement downwards of facial structures by the enlarging frontal lobes.

Even in *Lemur*, which retains a relatively elongated dog-like muzzle, the elongation takes on a downward trend compared with that of a dog. The angle of kyphosis, between the basicranial and basifacial axis is now 128° compared with 150° in the dog. In short-faced lemurs and the lorisoids (*Loris, Nycticebus, Perodicticus*) it is still more acute, whilst in Old World monkeys and hominoids, even when secondary elongation of the muzzle takes place (e.g. in *Papio, Mandrillus* and *Theropithecus*), the angle approaches 90° (see especially Hofer (1954, 1957, 1960, 1969a)). It follows that in *Lemur* and the long-snouted simians there is no interference with vision, for the axes of the frontally directed eyes are unimpeded, since the muzzle is displaced downwards beneath them.

11

The Evidence of the Teeth

Teeth are important, not only in assessing the relationships of existing primates, but also because on account of their durability they are often the only available remains of fossil forms. Although isolated aberrations of dental morphology occur here and there posing problems of their own yet, on the whole, the evidences of systematic affinity provided by the teeth are fully consistent with evidence from other sources. Hence inferences drawn from tooth structure is regarded as reliable by students of evolution. Dental structures betray little tendency of rapid adaptation to changed circumstances; their principal features, at any rate, being genetically determined. This is proved by the experiments of Glasstone (1938) who explanted undeveloped and undifferentiated molar tooth germs of rats and rabbits to tissue culture, in complete isolation from their normal environment. They developed in the artificial medium showing their characteristic crown pattern, with the cusps normal in number, shape and relations.

A heterodont dentition (i.e. one in which teeth differ in their form and functions according to their position in the jaws) has been inherited by the mammals from their theromorph reptilian ancestors, in which such dental differentiation was first to appear. Another feature, inherited from the reptiles and through them from even earlier vertebrate types, is the phenomenon of tooth succession during the life of the individual. Although continuous replacement of worn teeth by newly formed successors was the basic phenomenon (e.g. in sharks), this process has been limited in mammals to a diphyodont condition where a primary dental series (milk or deciduous dentition) is gradually replaced, during the growing period, by a second or permanent set. Primates are no exception to this rule and in them both dentitions are heterodont; the anterior teeth (incisors) being adapted for biting and cutting, followed in each jaw by a single tooth on each side that is elongated and its conical crown adapted for piercing. Beyond the canines come the cheek-teeth, generally further differentiated into molars or grinders at the rear and premolars in front, the latter generally showing transitional characters between the simple conical type of crown of the canines and the more complicated cuboidal crowns, adorned with cusps or tubercles on their chewing surfaces, of the true molars.

The number of representatives of these tooth categories varies in different mammalian (including primate) genera. It is customary to express these variations in a dental formula, the numbers of each category being indicated for half of each jaw above and below, sometimes with the initials I, C, P, M before the appropriate number (for the deciduous dentition the initial D. is prefixed before each category).

The basic eutherian mammalian permanent dentition is believed to be represented by the formula $\frac{3.1.4.3.}{3.1.4.3.} = 44$; a complement wherein the individual teeth would be designated: I_1^1; I_2^2; I_3^3; C_1^1; P_1^1; P_2^2; P_3^3; P_4^4; M_1^1; M_2^2; M_3^3. This notation is important insofar as loss of any member does not alter the designation of the remainder; e.g. when, as in man, only two premolars remain, these are $P_{\bar{3}}^{3}$ and $P_{\bar{4}}^{4}$ not, as might be supposed, $P_{\bar{1}}^{1}$ and $P_{\bar{2}}^{2}$. The scheme is a morphological system, not a mere numerical one.

In no living primate nor in any of the known fossil species is the primitive dental formula retained. None have retained 3 incisors in the upper jaw and the only family which retains 3 lower incisors is the Tupaiidae. Some fossil lemurs (e.g. *Adapis*, *Notharctus*) have retained the full number of premolars, but there has been some reduction by one in modern lemurs and New World monkeys (Cebidae) and a further reduction to two in all Old World monkeys and apes. Loss of premolars commences with $P_{\bar{1}}^{1}$, followed by $P_{\bar{2}}^{2}$ and appears to be correlated with the shortening of the jaws.

As regards molars, the full number is commonly retained, but the third molar ($M_{\bar{3}}^{3}$) is regularly absent in the marmosets and tamarins (family Hapalidae or Callithricidae) and as a frequent anomaly in *Hylobates* and man. Though retained, this tooth shows retrogressive changes in the lemuroid family Indriidae, in many New World genera and in the chimpanzee, while $M^{\bar{3}}$ only is so affected in *Tupaia* and *Perodicticus*. On the contrary, in long-faced forms such as baboons, enlargement of $M_{\bar{3}}^{3}$ is manifested, especially an antero-posterior lengthening of $M_{\overline{3}}$, which frequently develops extra cusps. The same tendency is shown in *Pongo* and *Gorilla*, $M_{\overline{3}}$ being larger than $M_{\overline{2}}$, whilst *Pongo* often develops supernumerary molars.

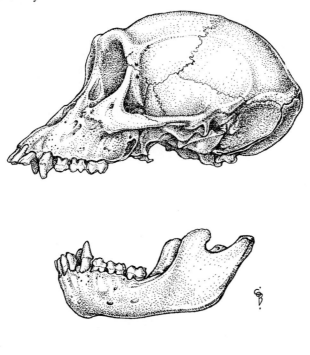

30 mm

FIG. 36. Left lateral view of the cranium and mandible of an immature chimpanzee (*Pan troglodytes*). The lower deciduous canine is about to be shed, DM. $_{\overline{1}}$ and $_{\overline{2}}$ still in situ and M. $_{\overline{1}}$ has erupted.

Incisors

Possibly, the numerous different roles the front teeth are liable to be called upon to perform (other than the mere seizing and cropping of food) may explain the diversity of their morphology among the primates. Primitively they are vertically implanted and exhibit simple peg-like crowns. The upper incisors of prosimians retain these features but their lower counterparts are grossly modified to serve, with the lower canines, as a toilet apparatus used in combing the fur, either in self-grooming or in social grooming. For

this purpose all the lower front teeth, including the canines, are inclined forwards (procumbent), elongated and the crowns laterally
compressed with pointed tips. These modifications had not occurred
in the fossil Eocene lemurs (*Notharctus*, *Adapis*). Extreme specializations have occurred in the front teeth of the aberrant aye-aye
(*Daubentonia*), a strange nocturnal lemur of Madagascar, which has
the unique dental formula $\frac{1 \cdot 0 \cdot 1 \cdot 3 \cdot}{1 \cdot 0 \cdot 0 \cdot 3 \cdot} = 18$. The incisors are elongated
and have assumed a form and growth pattern reminiscent of the
rodents; they are utilized in gnawing bark and wood in search of
the grubs of wood-boring beetles which form the animals' staple
diet.

In *Tarsius* (D.F. $\frac{2 \cdot 1 \cdot 3 \cdot 3 \cdot}{1 \cdot 1 \cdot 3 \cdot 3 \cdot}$) upper incisors are vertically implanted
and exhibit simple conical crowns; lower incisors (reduced to a
single tooth each side) show no more than slight procumbency –
less, in fact, than in some of its fossil forebears (e.g. *Omomys*).

Procumbency of the lower incisors in moderate degree recurs in
several simian genera, notably of the cebid sub-family Pitheciinae
(*Pithecia, Chiropotes, Cacajao*). Lower incisor crowns are specially

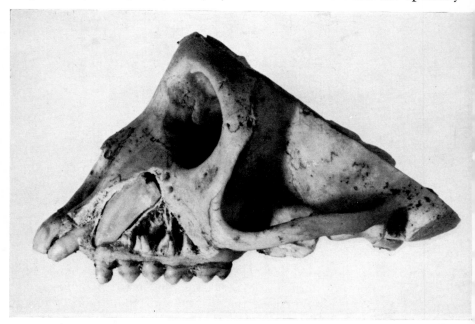

FIG. 37. Cranium of an immature male Rhesus monkey (*Macaca mulatta*)
with the surface bone dissected away from the maxilla to expose the
unerupted developing upper canine tooth, lying above the deciduous
canine, and roots of the premolars and two molars. × $\frac{1}{1}$

lengthened in one group of genera of the Hapalidae (Callithricidae).

Canines

These are extremely variable teeth, both as regards size, shape, degree of projection of the crown beyond those of neighbouring teeth and manner of occlusion with the teeth in the opposing jaw. For diagnostic purposes it is well to bear in mind that, although the morphology may be misleading, the lower canine interlocks in front of the crown of the upper canine, which itself occludes with that of the foremost lower premolar.

The upper canine crown is typically larger than the lower – even in man, whereas the lower does not project beyond the level of the incisors. The degree of canine projection is reflected in the size of the gaps left in the opposing dental series for the reception of the crowns during occlusion. These gaps or diastemata are between lateral incisor and canine in the upper jaw; between canine and premolars in the mandible. Stoutness of the canine crown varies much in South American monkeys e.g. they are long and trenchant in tamarins and in *Chiropotes*, whereas in *Callicebus* they project but little. The most elongated and trenchant canines occur in the terrestrial forms of Old World monkeys (baboons, drills, mandrills and geladas), where they also exhibit extreme sexual dimorphism. The gibbons show elongated canines in both sexes. Most robust canines of all occur in the great apes, where also sexual dimorphism is noted. Functional implications of these diversities are dealt with in Napier and Napier (1967).

Cheek Teeth

Two evolutionary trends appear to be manifested in the cheek teeth of primates – a reduction in the number of elements and an increase in the number of cusps adorning the occlusal surface of the crowns. Numerical reduction affects mostly the premolars, which are suppressed from the front, $P\frac{1}{1}$ being the one commonly lost, as in *Lemur* and *Tarsius*. In the New World families also $P\frac{2}{2}$, $P\frac{3}{3}$ and $P\frac{4}{4}$ are always present; but in the marmosets and tamarins there are but two molars. In all Old World primates $P\frac{3}{3}$ and $P\frac{4}{4}$ alone remain, with all three true molars.

In general, premolars in prosimians are simple, conical (i.e. unicuspid) teeth but the foremost is modified to resemble a canine and to function as such, while the hindmost varies in the direction of molar morphology (molarization), especially in Tupaiidae and the lorisoids, though not in *Lemur* and its allies nor in *Tarsius*. In Hominidae, both fossil and recent, upper and lower premolars are bicuspid; but in anthropoid apes the upper premolars are bicuspid and so is $P_{\overline{4}}$ but $P_{\overline{3}}$ is unicuspid (proterocone), the secondary cusp (deuterocone), if represented, being little more than a subsidiary swelling on the inner (or lingual) side of the main cusp. Thus $P_{\overline{3}}$ has a sectorial function as in Old World monkeys, a feature which modern apes also share with all known Miocene and Pliocene apes, but not with fossil or recent man.

The crowns of the upper molars are elaborated upon the basic primitive eutherian pattern of trituberculy, i.e. one in which three main cusps are arranged in a triangle (trigone) on a raised area, with a depressed talon (talon basin) lying posteriorly. Of the three cusps, one (the protocone) lies to the lingual (or palatal) side of the crown, the other two on the buccal side, the paracone in front and the metacone behind. An enamel thickening (cingulum) of variable development surrounds the bases of the cusps and is frequently the site of additional cuspules (styles). The upper molars of *Tupaia* and of *Tarsius* represent this, the most primitive molar pattern in the Primates. In *Tupaia* the mesostyle is particularly well marked and commonly bifid, whereas in *Ptilocercus* the mesostyle is absent but parastyle and metastyle indicated.

In all other primates there is a progressive tendency towards the assumption of a quadritubercular form by the uprising of a fourth main cusp – a *hypocone*. The same result may be brought about by either of two processes. Most commonly the hypocone is an upgrowth from the internal cingulum at the postero-medial angle of the crown. In advanced forms it attains the size of the main cusps. In other cases a *pseudo-hypocone* develops by splitting off from the protocone. This arrangement occurs in the Eocene fossil lemurs (subfamily Notharctinae) and sharply distinguishes them from the contemporary European forms (Adapinae) where the fourth cusp is a true hypocone.

Several stages in the transition from the tritubercular to a fully quadritubercular upper molar are witnessed in the present day lemuroids. In the dwarf lemurs (Cheirogaleinae) tritubercular upper molars persist; in *Lemur* and *Hapalemur* an incipient hypocone is found; a hypocone is fully developed in $M_{\overline{2}}$ of *Lepilemur* and *Phaner*, whereas in the indrises fully quadritubercular teeth are present. The lorises too are provided with four main cusps, though the hypocone varies in size in the different genera, and there are several supplementary cuspules.

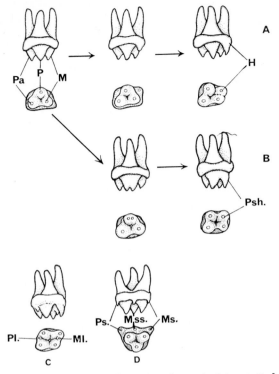

Fig. 38. Diagrams to illustrate the way the primitive tritubercular upper molar can be transformed into a quadritubercular tooth, due to the development of a true hypocone arising from the cingulum at the base of the tooth (A) or a pseudohypocone, formed by fission of the protocone (B). Side view of tooth shows medial aspect. As in C. and D. a further elaboration may occur wherein the crown of the upper molar may produce additional cusps, protoconule and metaconule, or styles, arising from the external cingulum.

H.	hypocone	P.	protocone
M.	metacone	Pa.	paracone
Ml.	metaconule	Pl.	protoconule
Ms.	metastyle	Ps.	parastyle
Mss.	mesostyle	Psh.	pseudohypocone.

At this point it is convenient to discuss the notation and nomenclature of additional or supplementary cusps that frequently complicate the crown patterns of upper molars. Enamel ridges termed lophs are often present connecting the protocone and paracone (*protoloph*) and metacone with hypocone (*metaloph*). Upon these, accessory cuspules, protoconule and metaconule are sometimes developed. Other accessory cuspules frequently appear on the buccal cingulum; these are known as *styles* and labelled *parastyle, mesostyle* and *metastyle,* according to their position, from the front backwards. They are well developed in lipotyphlan insectivores.

Among the simian families only the marmosets and tamarins (Hapalidae) retain tritubercular upper molars. All the remainder bear quadritubercular or quinquetubercular molars. The fifth cusp in this event is most likely to be found on $M^{\underline{3}}$. It is termed a *hypoconule* and is an enlargement of the rim of the talon basin. It is well marked on $M^{\underline{3}}$ in macaques, mangabeys and baboons. A peculiar specialization of the molars of Old World monkeys and not seen in the apes is the presence of high transverse lophs connecting the buccal and lingual cusps in pairs. The bilophodonty is seen in its most exaggerated form in the tribe Theropithecini, a group represented by the gelada and its extinct allies (e.g. *Simopithecus*) where the cusps and their connecting ridges are very elevated and the intervening valleys steep. In contrast, the anthropoid apes exhibit upper molars where the protocone and metacone are connected obliquely by a ridge with consequent formation of antero-lateral and postero-medial depressions or *foveae*. A similar arrangement occurs in some New World genera (*Ateles, Alouatta*). Occasionally in gibbons, but rarely or never in the great apes, a *tuberculum anomalum* (Carabelli's cusp) appears as an upgrowth from the cingulum at the mesio-lingual angle of the crown of $M^{\underline{1}}$ or $M^{\underline{2}}$. In man, it occurs in Neanderthal teeth and is common in Negroids, Asiatics, Malagasy, Micronesians, Polynesians and Amerinds. It also appears commonly on D. $M^{\underline{2}}$ in modern types.

The basic structure of a lower molar crown is described as tuberculo-sectorial or tribosphenic. The occlusal surface is divided into two parts, a front moiety bearing three cusps arranged in a triangle (the *trigonid*) and a rear section, on a lower plane, termed the *talonid* or talonid basin. Of the cusps on the trigonid, one, the *protoconid,* is located on the buccal side and the other two, *paraconid* and *metaconid,* on the lingual side. The talonid basin serves for the reception of the protocone of the corresponding upper tooth, when the jaws are closed. From its rim are developed one cusp (*entoconid*) on the lingual and another (*hypoconid*) on the buccal side, so that the total complement of a primitive primate lower molar is 5 cusps.

Evolutionary trends in lower molar crowns are towards the establishment of a quadritubercular pattern. This is accomplished by the gradual reduction and, finally, the disappearance of the paraconid, combined with a gradual shift of the protoconid until it lies more nearly opposite the metaconid. At the same time the previously narrow talonid widens and its level is eventually raised to that of the trigonid, the basin fills up and the two cusps, entoconid and hypoconid, enlarge to the size of the protoconid and metaconid. In many cases, notably on M$_{\overline{3}}$, a fifth cusp (*hypoconulid*) is developed on the posterior rim of the talonid basin, rather more to the buccal than the lingual side. Medial to it, in a few higher primates (e.g. *Gorilla* and the fossil *Dryopithecus*), a *tuberculum sextum* is also found.

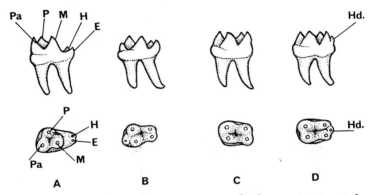

Fig. 39. Diagrams to show the manner in which a primitive tuberculo-sectorial lower molar may become converted into a quadritubercular tooth (A). B. shows the paraconid shrinking, while the entoconid and hypoconid are being elevated. In C. the disappearance of the paraconid is displayed while the other remaining cusps become subequal in size. D. shows the additional cusp, hypoconulid. Side view of teeth are shown from the mesial side.

E.   entoconid   M.   metaconid
H.   hypoconid   P.   protoconid
Hd. hypoconulid   Pa. paraconid

As in the upper molars accessory cusps (*stylids*) frequently make their appearance on the buccal cingulum, whilst crests (*lophids*) recalling those on the upper molars also appear, especially in the Cercopithecidae.

In *Tupaia* and other tree-shrews the lower molars retain the primitive tribosphenic character, with a raised trigonid bearing 3 cusps and a depressed talonid with entoconid and hypoconid and with a hypoconulid on M$_{\overline{3}}$. The same picture is also seen in the lower molars of *Tarsius*. The paraconid, however, was lost on the molars of the extinct Eocene genera *Necrolemur* and *Microchoerus* (of the subfamily Necrolemurinae) which had already attained a flat occlusal surface by raising the talonid. These genera also present progressive features in their upper molars not retained in their surviving representative, but which nevertheless adumbrate the conditions met with in the simian primates, a transitional stage being seen in the dentition of the important Oligocene fossil primate, *Parapithecus,* from the Fayûm deposits in Egypt. In this form, too, the paraconid had completely disappeared and a basically quadrituberculate molar attained with metaconid, protoconid, entoconid and hypoconid as well as a relatively well-developed hypoconulid.

In the Lemuroidea the lower molars show laterally compressed crowns and a division between the trigonid and talonid. Various stages in the elevation of the talonid are encountered in tracing the evolution of lower molars from the conditions seen in the earliest Notharctinae to those in Recent genera. In the earliest phases, the paraconid is still evident, at least on $M_{\overline{1}}$ and $M_{\overline{2}}$, while the hypoconid is connected by a *crista obliqua* with the posterior wall of the talonid basin. This results in the presentation of a double V pattern, a smaller V anteriorly, formed by the trigonid cusps and a larger V behind formed by hypoconid, entoconid and crista obliqua (Gregory, 1920b, 1921). Both in fossil and recent lemurs a hypoconulid is present marginally on $M_{\overline{3}}$.

In the lorisoids the lower molars are quadrituberculate; protoconid and metaconid lie opposite each other, while the talonid basin shows an entoconid and hypoconid, though these are more constricted than in lemurs. All four cusps are equal in size on $M_{\overline{1}}$ and $M_{\overline{2}}$, but $M_{\overline{3}}$ tends to be small due to reduction of the talonid; though in *Loris, Nycticebus* and *Arctocebus* all the cusps are present as well as a hypoconulid. In *Galago* the talonid basin of $M_{\overline{1}}$ is wide and shallow, with its cusps obsolescent.

In the Upper Eocene *Amphipithecus* from Burma, the dental formula is the same as in the South American monkeys; it was evidently a very early simian, though $P_{\overline{2}}$ was becoming vestigial. On $M_{\overline{1}}$ the tuberculo-sectorial character had been lost, talonid and trigonid being at the same level.

In the marmosets the loss of $M\frac{3}{3}$ is an indication of specialization, for these teeth are retained in the other New World monkeys, though showing some regression, especially in spider monkeys (*Ateles*). All lower molars are quadricuspid and the anterior pair usually joined by a transverse lophid, the paraconid and hypoconulid have disappeared, though traces remain in *Alouatta* and *Ateles.*

In Old World monkeys lower molars have been specialized in the same way as their upper opponents, i.e. they are bilophodont, two lophids or enamel ridges connecting the anterior and posterior pairs of cusps transversely. A hypoconulid occurs on $M_{\overline{3}}$ in *Macaca* and in baboons and mangabeys, located on a heel-like projection of the talonid. This is marked off by grooves on the buccal and lingual aspects of the crown; similar grooves separate the trigonid and talonid moieties of each lower molar. Fossil evidence indicates that bilophodonty in Old World monkeys had been established at least by the beginning of the Miocene.

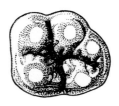

A   B

FIG. 40. Occlusal surface of the crown of the molars of a baboon (A) and a chimpanzee (B) to show contrasts in the cusp pattern etc. Note the bilophodont character of A. and the similarity in B. to that of *Dryopithecus* (Fig. 44, p. 87). Redrawn from Simons.

On the other hand, dental differentiation of the anthropoid apes from the Old World monkeys seems to have become established in the Oligocene. At this period, on the evidence of fossils from the Fayûm deposits in Egypt, ancestral gibbons (*Propliopithecus, Aegyptopithecus, Aeolopithecus*) had already made their appearance; the first mentioned, *Propliopithecus,* being perhaps at the point of division between the lineage leading to man and that which produced the first large apes (*Dryopithecus*) (Simons, 1965). In none of these does bilophodonty appear; instead $M_{\overline{1}}$ and $M_{\overline{2}}$ show rounded contours with low, rounded cusps with entoconid and hypoconid joined by a distinct crest. M tends to be reduced.

In the dryopithecines (*Dryopithecus, Sivapithecus, Proconsul*) which flourished during the Miocene and Pliocene, a special pattern of grooves called the Y.5 pattern (Hellman, 1928) is manifested. The details of the Y-shaped arrangement have been frequently misinterpreted but, according to Robinson and Allin (1966), the basic or typical pattern is one in which the hypoconid occupies the space between the bifurcated limbs of the Y; the stem of the Y separates the metaconid and entoconid while two short antero–posteriorly aligned grooves separate protoconid from metaconid anteriorly and entoconid from hypoconid posteriorly. The anterior fissure terminates in the well developed anterior fovea. The posterior fissure, which starts at the base of the distobuccal limb of the Y, terminates in the posterior fovea, which is typically not so well marked as the anterior.

From this so-called Dryopithecus pattern, by modification of the fissure pattern and with or without cusp reduction, a number of variations are derived and are found in the dentition of the present day apes and man. Man is characterized by one of these modifications, wherein the hypoconulid is suppressed, thus giving a +4 pattern (see also Riethe (1967)).

6 mm

FIG. 41. Occlusal surface of crown of an upper molar of a Green Monkey (*Cercopithecus sabaeus*) to show its bilophodont character.

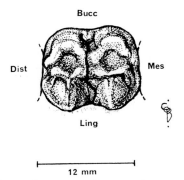

Bucc

Dist

Mes

Ling

12 mm

FIG. 42. Occlusal surface of crown of a partly worn left lower second molar of a female baboon (*Papio hamadryas*) to show the arrangement of cusps. Bucc. buccal; Ling. lingual; Dist. distal; Mes. mesial.

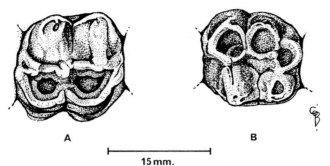

A   B

|—————————————————|
15 mm.

FIG. 43. Occlusal surface of the crown of a left upper (A) and a right lower molar (B) of an adult female Gorilla, showing patterns of wear on the cusps.

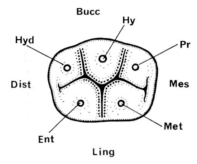

FIG. 44. Left lower molar of *Dryopithecus*, with the Y pattern of grooves on the occlusal surface shown diagrammatically. Redrawn from Robinson and Allin (1966). Bucc. buccal; Ling. lingual; Mes. mesial; Dist. distal; Ent. entoconid; Hy. hypoconid; Hyd. hypoconulid; Met. metaconid; Pr. protoconid.

D

12

Diet and Digestion

The earliest mammals scurrying about the forest floor at night were limited dietetically to what they could find among dead leaves, moss and between stones, i.e. an almost purely insectivorous diet. The present day lipotyphlan insectivores carry on this regime with a tendency, in the case of the larger members (tenrecs, hedge-hogs), to become omnivorous as in the case of their marsupial counterparts, the opossums. These are persistent scavengers, in-cluding in their regimen almost anything edible, whether of animal or vegetable origin.

With the adoption of an arboreal habitat and diurnal activity a change of diet to a basically vegetable regimen was advantageous though resort, in some measure, to animal food was retained in order to supplement the protein intake, since vegetable protein is low in biological value.

The forest environment produces an abundance of vegetable food in the way of leaves, shoots, bark, pith, seeds, nuts and fruits as well as roots and tubers. Some primates rely wholly on these aliments, but most supplement with some item of animal origin in the way of insects, molluscs, crustacea, spiders and, on occasion, foods of vertebrate origin, lizards, geckoes, tree-frogs, birds and their eggs and even smaller mammals.

Tree-shrews are largely insectivorous, their food consisting of insects and earthworms, though some fruit is ingested; but *Urogale* adds lizards, young birds and birds to its fare (Wharton, 1950). The lorisoid prosimians, including the galagos, fall into the same dietetic category as do also the small nocturnal lemuroids such as

Microcebus, Cheirogaleus, Lepilemur and also *Tarsius.* On the contrary, the larger diurnal lemurs are more purely frugivorous.

Relatively few primates have become specialized to a highly restricted diet and none to the extent found in such mammals as the koala or the giant panda. One group, however, in each of the main systematic divisions has become adapted ecologically and morphologically to a purely leaf-eating (phyllophagous) habit. The family Indriidae, among the Malagasy lemurs, the platyrrhine genus *Alouatta* (howler monkeys) and, among the Old World monkeys, the whole family Colobidae (the African *Colobus* monkeys and the Asiatic leaf-monkeys) are in this category.

The platyrrhine subfamily Pitheciinae (*Pithecia, Chiropotes, Cacajao*) are reported to be purely frugivorous in the wild (Fooden, 1964).

Another specialized feeder is the aye-aye (*Daubentonia*), which shows a highly specialized rodent-like dentition and a modified narrow, elongated wire-like middle finger. These are adaptations to the habit of gnawing bark to expose the burrows of wood-boring beetle-grubs which are then enucleated with the modified finger and then consumed.

Most Old World monkeys as well as the gibbons and the great apes are basically frugivorous, but some insect or similar supplementation commonly occurs in the wild. In those forms which have successfully exploited the savannah (*Erythrocebus, Papio*), grasses (including their underground stems and roots) are added to the menu and in the baboons a secondarily carnivorous habit is occasionally manifested (Dart, 1963; DeVore and Washburn, 1963; DeVore and Hall, 1965).

The heavily endowed masticatory apparatus of the gorilla seems to be correlated with a purely vegetarian diet, its food in the wild state being composed largely of wild celery, bamboo shoots and young shoots of giant lobelias and *Senecio* (Schaller, 1963), though lowland gorillas eat some fruit.

Morphological Adaptations

The comparative visceral anatomy of surviving primates has much to teach us regarding the phases and progress of primate evolution, in spite of the lack of collateral evidence from paleontology – apart from such as is to be inferred from the dentition of fossil forms (see previous Chapter).

COMPARATIVE LINGUAL ANATOMY

The tongue, of course, is the chief locus of the sense of taste, insofar as its epithelial covering contains scattered receptor cells in sub-microscopic organs called taste-buds. As such it might appropriately have been considered with the receptor apparatus in Chapter 3. But the organ is concerned also in other aspects of alimentation, notably in dental toilet phenomena and in the act of swallowing (deglutition) as well as a modifier of vocal emissions and other forms of social communication. These functions are markedly reflected in its structure.

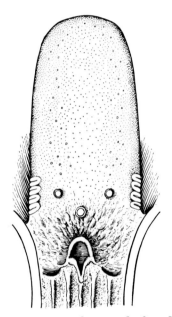

FIG. 45. Dorsal surface of the tongue of a Purple-faced Langur (*Kasi senex monticola*), showing distribution of fungiform and vallate papillae and the lateral organs. From Hill (1952).

Four kinds of papillae bedeck the epithelial surface of the tongue: conical, fungiform, vallate and foliate. Their details and distribution have been fully considered by Sonntag (1921). Conical or filiform papillae are most numerous, covering almost all of the dorsal surface. In the Lemuroidea, they extend to the pharyngeal part of the organ, where they are uniquely enlarged, specialized and thorny. These are also represented in Lorisoidea, extending back towards the epiglottis. They are absent in *Tarsius*, where also the filiform papillae are long and hairlike. Fungiform papillae are present sporadically among the filiform papillae over the anterior two thirds of the dorsum, but tend to be more numerous towards the apex and at the sides, with a few sometimes transgressing the lateral edge on to the ventral surface. The ventral papillary zone is wide in great apes and Old World monkeys, also in *Ateles, Lagothrix* and *Cebus,* but in all others narrow or absent.

Vallate papillae are few and large and usually arranged in a triangle with the apex posteriorly. Only in apes and in the Lemuridae are there more than four arranged in Y-formation. *Tarsius* is peculiar in having three arranged in a line.

Foliate papillae form the so-called lateral organs located on each side of the hinder-part of the oral section of the tongue, where they appear as a series of alternating clefts and ridges (laminae). Like the vallate papillae they are provided with taste buds. Lateral organs are absent in all lorisoids, *Microcebus, Daubentonia, Tarsius* and man. In the marmosets they are faint and irregular, with few laminae. In other platyrrhine monkeys they are ovoid with their medial borders concave. In Old World monkeys there are rows of short laminae and sulci along each lateral border. In gibbons, chimpanzees and all species of *Lemur* they resemble those of the Cebidae, but in the orang and gorilla they form a ladder-like arrangement on the dorsum with only their lateral ends cutting the lateral borders.

In the middle line of the free part of the tongue in primitive primates is a rod-like supporting structure called the lyssa (or lytta) composed partly of hyaline cartilage and partly of striated muscle-fibres. In *Nycticebus* these components are located in the space between the transverse and inferior longitudinal muscles, the cartilaginous element lying anterior to the muscular. The latter is a sausage-shaped mass encapsulated in connective tissue (Kubota and Iwamoto, 1967). In *Tupaia* the organ is wholly muscular and is regarded by Schneider (1958) as a support for the tongue. Bluntschli (1938) emphasized its supportive function in nocturnal primates where it is necessary for the tongue to be elevated to enable it to make tactile contact with food items.

Beneath the free anterior part of the tongue in tree-shrews and prosimians is another peculiar structure, the sublingua – a relic of the reptilian tongue and also well represented in marsupials and some other lowly mammals. In tree-shrews (especially in some Tupaiinae) the organ extends to the tongue apex and presents a serrated edge, but it is less specialized than in prosimians. In these the sublingua is a thin, flexible, horny plate with a median axial ridge (lyssa of the sublingua) on its under surface. Its edge is adorned with fine, styliform denticulations that adapt to the interdental intervals of the lower front teeth. They constitute the so-called dental comb. Observations on the living animal indicate that the sublingua functions as a tooth cleaner, removing hairs that have accumulated between the teeth during the autogrooming (toilet operations) or allogrooming for which the front teeth are themselves specialized. The lyssa of the sublingua is rod-like and composed of connective tissue only, but it is connected above to the lyssa of the tongue by adipose tissue and muscular fibres (Nusbaum and Markowski, 1896, 1897).

In *Tarsius* the sublingua is simple and somewhat fleshy, adherent at its apex, unserrated and only slightly free laterally. The same applies to the tongues of marmosets, where the small triangular sublingua has a crenulated margin. In all higher platyrrhines as well as in apes and man, the sublingua is reduced to a fold of mucous membrane, the plica fimbriata, and even this is lacking in many Old World monkeys.

A specialization in some New World monkeys (*Callicebus, Aotes*) has been referred to by Hofer (1969b) by the non-committal name of organon sublinguale, but erroneously labelled by earlier writers as sublingua or frenal lamella. Its homology is at present obscure, but histologically it contains no muscle fibres, so its movements depend on the tongue. It resembles a sublingua in presenting tentacle-like papillae along its edges, but these carry excretory ducts from salivary glands. The organ also bears structures resembling fungiform papillae and taste buds.

SALIVARY GLANDS

Representatives of all the major salivary glands recognized in lower mammals are to be found in Primates. These are the parotid, submandibular and greater (Bartholinian) and lesser (Rivinian) sublingual complexes (Huntington, 1913; Schneider, 1958). There are also found minor collections of glands in the lips, cheeks and palate variously developed.

In *Tupaia* the parotid is shaped like an inverted horseshoe around the auditory opening and its duct opens near the angle of the mouth, where it also receives the secretion of a small buccal gland. The submandibular gland is conspicuous. In *Ptilocercus* the sublingual gland is not present as a distinct entity, but represented by anterior lobules of the submandibular gland (Le Gros Clark, 1926). *Tupaia* would appear to be similar. The submandibular is the principal gland in *Lemur, Microcebus, Indri, Daubentonia, Galago* and *Tarsius* (Fahrenholz, 1937), but the parotid is well developed in *Lemur* and *Tarsius*.

The parotid is relatively largest in the platyrrhine douroucoulis (*Aotes*) but tends also to be hypertrophied, as do all the salivary glands, in specialized leaf-eaters such as *Alouatta* and the Colobidae, probably from the need to produce large quantities of mucus to assist deglutition of the bolus. In Old World primates isolated lobules (socia parotidis) develop along the course of the parotid duct. New World monkeys (other than *Aotes*) possess additional elements of the submandibular gland lying deep to the mylohyoid and some (*Ateles, Lagothrix*) have a well defined secondary gland, with its own duct besides accessory lobules appended to Wharton's duct. This secondary gland occurs also in some species of macaques (e.g. *M. nemestrina*).

Bartholin's gland is frequent but inconstant, especially in Old World monkeys. In New World forms it appears only in *Ateles* where it is inconstant, but independent of the secondary submandibular gland. Its drainage may be into Wharton's duct posterior to the parafrenular papilla (*Ateles, Pan*) or, more usually, by an independent duct opening on the papillae lateral and caudal to Wharton's duct (*Macaca, Papio, Pongo*).

The Rivinian complex (lesser sublingual gland) opens by a series of small ducts into the floor of the mouth, in the alveolo–lingual groove. The gland consists of a series of more or less discrete lobules and is invariably represented. It may be closely related to the Bartholinian gland, but there is no parenchymal continuity. In *Macaca* there are anterior and posterior segments connected by a narrow isthmus applied to the lateral aspect of the greater sublingual gland.

BUCCAL POUCHES

Buccal pouches are outpocketings of the cheek mucosa on either side from the lower alveolo–buccal groove opposite the premolar teeth. They are specializations developed only in one family (Cercopithecidae) of Old World monkeys and are absent in the other family (the leaf-eating Colobidae). Buccal pouches are invariably correlated with the possession of a simple, unmodified stomach. They vary in size, being small in *Cercopithecus*, large in macaques and baboons. Their function is to serve for temporary storage of hastily snatched food that can thereafter be masticated at leisure.

Though covered by layers of muscle fibres derived from the facial muscles, these are insufficient to empty the pouches without external manual assistance.

STOMACH

Compared with other mammalian orders, the alimentary tract in Primates betrays a relatively generalized structure, with little tendency to show extreme complications (e.g. rodents and ruminants) (Mitchell, 1905, 1916). This applies especially to the stomach, which on the whole, remains a simple globular or pyriform sac. The globular type, where the cardiac and pyloric openings are approximated, prevails in lorisoids, mouse-lemurs and marmosets; the pyriform characteristic of *Lemur* and most New World and Old World monkeys, also in the apes and man. Only in the specialized leaf-eaters is some gastric specialization encountered. This takes the form of enlargement of the corpus ventriculi and a tendency to sacculation, least marked in *Alouatta*, more advanced in the Indriidae, but most elaborate in the Colobidae. In the last mentioned the organ is greatly enlarged, displacing and distorting the liver. It is also divided into three parts folded upon one another, the first two parts being elaborately sacculated and provided, like the colon, with longitudinal muscles concentrated into two discrete bands (Owen, 1833, 1835; Flower and Lydekker, 1891; Ayer, 1948; Hill, 1952, 1958).

INTESTINE

The small intestine though varying somewhat in proportionate length is fairly uniform in its form and disposition throughout the primates but, due to specialization in the peritoneal relationships, the different regions are less distinctly defined in the lower types than in the more advanced. The simplest arrangement, as presented by the tree-shrews, *Tarsius*, some lemurs and *Saimiri*, is one where a simple bilaminar median dorsal mesentery suspends the intestine from the dorsal abdominal wall all the way from the pylorus to the rectum. The gut, being longer than the dorsal attachment, therefore hangs in a series of coils. Elsewhere anchoring takes place by a peritoneal thickening (cavo–duodenal ligament) at the aboral end of the duodenum. The duodenum then falls over to the right, becoming progressively fixed to the dorsal abdominal wall. The remainder of the small intestine then tends to occupy the left flank. Primitively the shortish large intestine accompanies it, lying dorsal to the small intestinal coils. In *Tupaia*, *Tarsius* and *Saimiri*,* the large intestine retains the form of a simple, straight tube with a short conical caecum (= caput caecum coli) at the cranial end adjacent to the ileo–colic junction.

In the more progressive majority of prosimians and all simians, the intestinal tract beyond the duodenal fixation undergoes rotations whereby the ileo–colic junction is carried to the right ventrally across the beginning of the small intestine. This implies longitudinal growth of the large intestine and eventually brings the ileo–colic junction and caecum into the region of the right iliac fossa, thus resulting in the large gut being disposed in the form of a horseshoe around the small intestinal coils. Some peritoneal changes are hereby involved. The mesentery of the proximal part of the colon no longer persists as a simple straight fold. In extreme cases, e.g. in man, the proximal mesocolon is lost by adhesion of its dextral leaflet to the parietal peritoneum of the right flank, the proximal colon becoming thereby anchored to the body wall. All stages of this process are represented as permanent conditions in the adults of one or other primate genera (Klaatsch, 1892b; van Loghem, 1903).

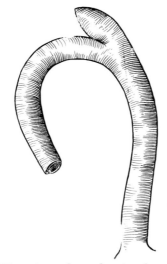

Fig. 46. Hind gut of *Tupaia* to show the simple pattern and cranial position of the caecum. Redrawn from Hill and Rewell (1948).

* Appearances in *Saimiri* are deceptive insofar as, though the colon and caecum are primitive, the peritoneal relations are not, being uniquely specialized for some unknown reason. In this genus the straight colon in late foetal life migrates from left to right, carrying its mesentery with it, the latter becoming stretched over the coils of small intestine, which thus become imprisoned in a peritoneal bag (*bursa mesocolica*) (Hill, 1960, 1969).

As to the colon itself, this shows numerous modifications that conform to the physiological needs, but also displays morphological evidence of genetic relationships.

In lipotyphlan insectivores the colon is a direct uninterrupted continuation of the small gut, i.e. there is no sphincter between them nor any blind outgrowth (caecum) at the junction. This simple condition is departed from in the menotyphlans, both the Macroscelididae and the Tupaiidae, in both of which a caecum is present. The simplest arrangement is presented by the pen-tailed treeshrews (*Ptilocercus*), where the short, straight colon of wide calibre passes from the ileo–colic junction to the anal canal suspended in a simple dorsal mesentery. At its cranial end a conical diverticulum, the caecum, projects forwards to the left of the ileo–caecal union. Already in *Tupaia*, a slight advance is shown insofar as the caecum bends over to the right taking the commencement of the colon with it, so as to indicate an incipient division of the latter into a transverse and a "descending" moiety. This trend is carried further in some specimens of *Tarsius*, where the caecum is longer and a well defined transverse colon is present. Very similar conditions prevail in the lowliest lemurs (*Microcebus* and *Cheirogaleus*).

In all other prosimians, both lorisoids and lemuroids, the transverse colon is more or less elaborately specialized by its elongation and by the formation of a U-shaped loop (ansa coli) the two limbs of which are bound together closely by a peritoneal fold. An additional peritoneal sheet (ansacolic ligament) in some cases anchors the proximal (oral) limb of the loop to the ascending colon when this has evolved by backward migration of the ileo–colic junction along the right flank. In *Loris* and *Arctocebus* the colic ansa is short and simple. In *Lemur* and *Daubentonia* it is longer. Further elongation and elaboration is seen in *Nycticebus, Perodicticus, Galago* and the Indriidae. In *Galago senegalensis* the loop is very long and bent over on itself to the left, whereas in *G. crassicaudatus, G. alleni* and *Galagoides demidovii* it turns to the right. In *Nycticebus* and the Indriidae the elongation is more extreme and the loop is accommodated by becoming rolled into a tight spiral – an exactly similar process to that occurring in ruminant Artiodactyla. In some cases, e.g. *Euoticus*, the colon is further differentiated by developing saccules separated by longitudinal muscle bands (taeniae coli).

In no simian primate is an ansa coli developed on the transverse colon, so this structure must be regarded as a prosimian specialization. The simple horseshoe shaped colon is characteristic of the marmosets and most of the other New World monkeys (other than *Saimiri*). The large gut, however, is relatively simple in *Alouatta*,

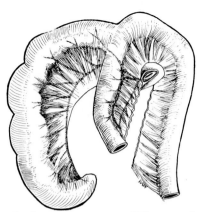

FIG. 47. Hind gut and related structures of *Tarsius*, from the ventral aspect. Based on Hill and Rewell (1948).

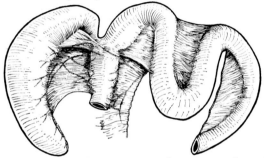

FIG. 48. Hind gut and related structures of *Loris tardigradus*, from the ventral aspect. Note the ansa coli and ansacolic ligament. Based on Hill and Rewell (1948).

though its commencement is grossly enlarged in calibre so as to resemble a second stomach – doubtless an adaptation to the bulky diet. Incipient sacculation is found in the colon of marmosets, but is lacking in *Cebus* and most other Cebidae, except *Ateles*, though fully developed taeniae coli with incipient sacculation occur also in *Lagothrix* and *Brachyteles*.

All Old World monkeys agree in having the colon markedly and permanently sacculated with three taeniae coli.

They are also characterized by having the colon well differentiated into ascending, transverse and descending parts. The first is rather short and relatively immobilized, the second elongated and arranged in a festoon across the belly and the third very long, thrown into numerous mobile coils and provided with a mesentery, continuous with that of the transverse colon.

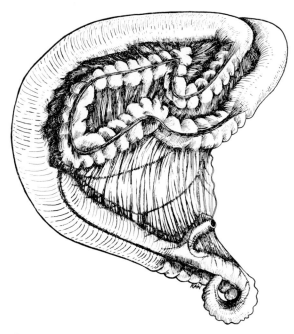

Fig. 49. Hind gut and related structures in the Potto (*Perodicticus potto*). Redrawn from Hill and Rewell (1948).

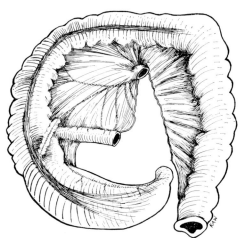

Fig. 50. Hind gut and related structures in the Common Marmoset (*Callithrix jacchus*). Redrawn from Hill and Rewell (1948).

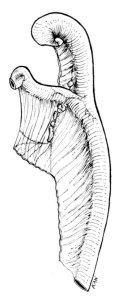

FIG. 51. Hind gut and related structures in the Squirrel monkey (*Saimiri sciureus*). Redrawn from Hill and Rewell (1948).

The apes and man differ in the greater degree of fixation of the ascending colon, whilst in man the descending colon has also become adherent to the abdominal wall, leaving free only the pelvic loop immediately preceding the rectum. In the great apes a peculiar reduplicature of the descending colon is formed by elongation producing a sigmoid pattern with the limbs closely apposed (Hill, 1949). In all the apes sacculation and taeniae affect the colon; in gibbons the taeniae are increased to four.

The caecum too follows to some extent the specializations of the colon. In menotyphlans it remains short and usually conical (e.g. in *Ptilocercus*), but it is more elongated, with a bluntly rounded apex, in *Anathana* and reduced in *Lyonogale* (= *Tupaia tana*). In *Tarsius* the caecum resembles that of *Anathana*, but is still longer, though of uniform calibre throughout. The same holds true for *Microcebus*, but here there is sometimes a basal enlargement. In all other prosimians some specialization occurs. In *Galago* and *Arctocebus* incipient sacculation is seen, but in the potto and *Avahi* the organ is permanently sacculated and provided with taeniae. *Hapalemur* has a short, wide hook-shaped caecum, whereas that of *Lepilemur* is of enormous capacity and coiled with a conical tip. *Lemur* and

Propithecus also have a relatively capacious, coiled caecum with narrowed apex.

In *Daubentonia* there is distinct differentiation of the caecum into a wide basal sac and and a terminal vermiform portion.

In the New World monkeys, including the Callithricidae, the caecum is typically short, wide and U-shaped and demarcated by a sphincter from the basal part of the colon, but in *Cebus* the linear axis is retained, the appearance being much as in *Tarsius,* while in *Alouatta* the sac is short and wide, partaking in the expansion of the basal colic sac already referred to. *Brachyteles* presents a similar picture, but retains the U, whilst *Ateles* is more typical.

In Old World monkeys, the caecum, though often curved to the left is never U-shaped. It shows distinct physiological demarcation into a thin walled basal sac, usually sacculated, and a thicker-walled apical portion, which may or may not be conical. Taeniae coli extend on to the basal sac, but spread out into a continuous muscular sheet on the apical region.

The gibbons, great apes and man agree in the characters of the caecum which is composed of a short rounded and sacculated

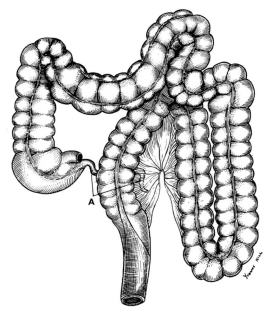

Fig. 52. Hind gut and related structures in an anthropoid ape (Orang-utan, *Pongo pygmaeus*) showing caecum, with appendix (A), the sacculated colon and the inverted, U-shaped loop on the hinder colon. From Hill (1949).

basal portion which narrows apically to constitute a morphologically differentiated vermiform appendix, characterized by its narrow lumen and thick wall in which lymphoid tissue is very prominent, recalling that region of the gut in the rabbit and its allies. Some minor differences in detail are seen in the different genera.

(For a fuller exposition consult Hill (1958) and for the caecum and its adnexa, Hill and Rewell (1948) and works there cited.)

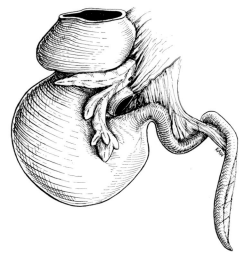

Fig. 53. Caecum and appendix vermiformis of Orang-utan (*Pongo pygmaeus*). Redrawn from Hill and Rewell (1948).

13

The Respiratory Tract

Upper Respiratory Tract (Nasal fossae)

With the gradual reduction in the size and projection of the muzzle, the passage from lower to higher Primates shows marked changes.

The tree-shrews and to a great extent the lemurs as well retain a primitive mammalian pattern of arrangements in the nasal fossae. The two fossae are separated by a vertical septum composed partly of bone, partly of cartilage, covered both sides with mucous membrane. Each lateral wall is embellished with a series of elongated inwardly growing scroll-like structures termed turbinate processes or conchae. These serve to increase the area of the mucous membrane, which being highly vascular, warms the inspired air before the latter gains admission to the lower tract. It also increases the olfactory area (see above, pp. 23–24) and there is some correlation between the complexity of the turbinates and olfactory acuity.

In lowly or macrosmatic mammals the conchae are numerous, with much secondary elaboration. In tree-shrews there are five main scrolls (endoturbinals): a naso-turbinal, growing from the nasal bone; a maxillo-turbinal, springing from the maxilla and four ethmo-turbinals numbered from the front backwards. In contrast to the complex maze formed from the maxillo-turbinal of rodents, carnivores and ungulates, that in the Tupaioidea and all Primates is a simple scroll. In tree-shrews, in addition to the five endo-turbinals there are two smaller ectoturbinals, one hidden beneath the naso-turbinal, between it and the first ethmoturbinal, and the second between ethmoturbinals 2 and 3.

The same general pattern as regards number and disposition of the conchae is found in lemurs, with the exception of the aye-aye (*Daubentonia*), where the system is less reduced, indicative of a more primitive status and greater dependence on the sense of smell.

A definite specialization occurs in the lorisoids where the first ethmo-turbinal is extremely large, extending below to cover the maxillo-turbinal.

In *Tarsius* a process of reduction and simplification commences. Both naso-turbinal and maxillo-turbinal are shortened pari passu with the shortening of the muzzle and ethmo-turbinals are reduced to two, the postero–superior two having been lost.

The nasal chambers in all higher Primates, from marmosets to man, are reduced in size and turbinals share in this. The naso-turbinal becomes a mere ridge or thickening near the roof and the maxillo-turbinal a simple scroll. Only two, occasionally three, ethmo-turbinals remain and these become rearranged so as to lie one above the other instead of the earlier fore and aft arrangement. The first ethmo-turbinal remains the largest and hides a single ecto-turbinal which persists as the bulla ethmoidalis.

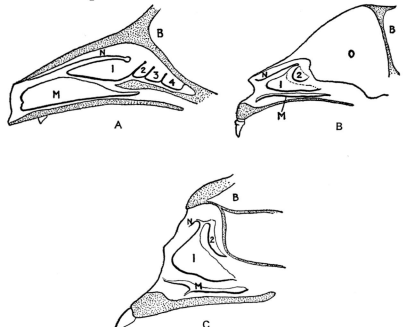

FIG. 54. Paramedian sagittal section of the right nasal fossa, seen from the lateral surface in A. *Lemur*; B. *Tarsius* and C. *Callithrix*.
B. brain cavity   N. nasoturbinal   1–4 ethmo-turbinals
M. maxillo-turbinal   O. orbit

The highest Primates (man and anthropoid apes) are remarkable for the number and complexity of the so-called accessory nasal *air-sinuses*, outgrowths from the nasal mucosa invading various cranial bones, lightening them and also modifying the voice. The largest and phylogenetically the oldest of these air spaces is the maxillary air-sinus or antrum. This is an outgrowth from the neighbourhood of the bulla ethmoidalis into the body of the maxilla. Other outgrowths invade the frontal, presphenoid and ethmoid bones. Only the chimpanzee and gorilla show an arrangement of frontal and ethmoidal sinuses comparable to that of man (Keith (1921) but see also Wood Jones (1939) and Cave and Haines (1940)).

In Old World monkeys the maxillary sinus is usually the only accessory sinus present (Weinert, 1925) but a small frontal sinus has been reported in some examples of *Cercopithecus* (Aubert, 1929; Hill, 1966). In *Papio* even the maxillary is absent due to the amount of space occupied by the large canine root.

In the Cebidae, *Ateles* presents a small vesicle only in the maxilla and in addition distinct frontal sinuses, but none in the presphenoid. *Alouatta* is similar, but also develops sphenoidal sinuses and has larger maxillary air spaces (Wegner, 1956). *Cebus* also presents frontal sinuses; these, however, may not be homologous with those of man (Cave and Haines (1940) though Wegner (1956) disagrees). Sphenoidal sinuses are likewise present in *Cebus*.

The prosimians appear to have developed a system of air-spaces independently from the simians, albeit affecting the same parts of the skull. Thus in the Lorisoidea frontal sinuses appear in *Loris*, but not in *Nycticebus*, *Arctocebus* or *Perodicticus*, but a large maxillary antrum is present in all (Kollmann and Papin, 1925). Galagos are stated by the same authors to resemble the Lorisidae.

In *Lemur* a remarkable diversity affects the sinuses, which differ in arrangement among the five species of the genus. Large frontal and maxillary sinuses are present, the latter sometimes subdivided into an upper and a lower portion. Palatine sinuses are developed in *L. mongoz* and *L. fulvus*, but not in *L. macaco*, *L. catta* or *L. variegatus*. Sphenoidal and ethmoidal sinuses are also variably developed, being present in *L. mongoz* and in the Indriidae (for details, see Kollman and Papin (1925)).

Larynx and Air Sacs

Basically the structure of the primate larynx conforms to that of mammals generally, showing little variation in the disposition of its cartilaginous components or of the mucous lining, irrespective of the vocal sounds emitted.

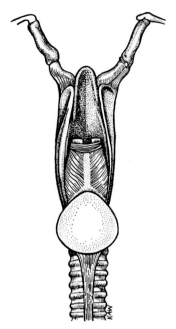

FIG. 55. One form of dorsal median air-sac as shown in the Indris (*Indri*), a large lemuroid. Redrawn from Starck after Milne-Edwards.

In primitive (and indeed all macrosmatic) mammals, which move about with their noses to the ground, the epiglottis is large and projects upwards dorsal to the elongated soft palate. As a result, breathing and olfaction can proceed unhindered even when the animal is feeding or has its mouth open for any other reason. This arrangement is still effective in many subhuman Primates such as the prosimians, marmosets and even such types as *Cebus* (Negus, 1949). In Man and other higher Primates this arrangement is lost for, although the epiglottis is still large, it is too short to ride over the shortened soft palate so that the buccal cavity cannot be completely shut off. It therefore becomes easier to breathe through the mouth than the nose – a situation of doubtful biological advantage. Moreover, in order to appreciate odours, it becomes necessary to keep the mouth closed and take a deep sniff.

Greatest variability in laryngeal structures resides, however, in the air sacs – laryngeal appendages that have evolved independ-

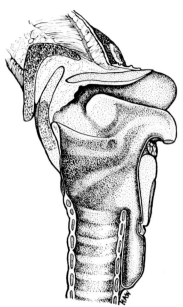

FIG. 56. Dorsal median sagittal view through the larynx of a Spider monkey (*Ateles*) showing especially the dorsal air-sac. Redrawn from Starck.

ently in different groups from different locations. These have been summarized by Starck and Schneider (1960).

Median outgrowths may develop from the ventral or from the dorsal aspect of the respiratory passage. They enlarge sufficiently to burrow beneath adjacent neck or throat muscles which can be used to compress them.

A dorsal outgrowth (*saccus laryngo–trachealis posterior* of Bartels) escapes from between the cricoid cartilage and the first tracheal ring extending thence caudally between trachea and gullet and has so far been recorded only in *Indri, Lemur variegatus, Microcebus, Ateles* and *Alouatta* – in the last mentioned in combination with other sacs.

Ventral outgrowths take place from two different positions but do not coexist. Commonest is the *saccus laryngeus medius superior* or subhyoid air sac. This is an unpaired outgrowth which escapes through a T-shaped hole at the base of the epiglottis and reaches the exterior of the larynx between the thyroid cartilage and the hyoid bone, thereafter insinuating itself between the infrahyoid muscles. According to Lampert (1926), this sac is well developed in the platyrrhine genera *Lagothrix* and *Alouatta,* but is lacking in

Ateles, Cebus, Saimiri, Pithecia, Aotes and Callithricidae. In *Alouatta* this sac is greatly hypertrophied, where it invades the hyoid bone, inflating it into a drum-like structure which is responsible for the volume and carrying power of the voice. Minor variations in this bone affect the different species of the genus.

The superior laryngeal sac is also present in many Old World monkeys, though lacking in the Hominoidea. It is well developed in *Cercopithecus, Cercocebus, Macaca, Papio* and in the Colobidae (except in *Procolobus*). It spreads over the front of the neck as far as the shoulders and likewise forwards over the hyoid (e.g. in *Macaca nemestrina*).

An inferior laryngeal sac is rare among Primates, though frequent in rodents and equines. Such an outgrowth escapes between the thyroid and cricoid cartilages by perforating the crico–thyroid membrane and occurs only in marmosets and tamarins.

Paired lateral outgrowths that arise as diverticula from the laryngeal ventricle (the space between the false and true vocal cords) are characteristic of the Hominoidea. Escaping through the thin lateral parts of the thyrohyoid membrane, they enlarge beneath the superficial muscles of the neck, spreading in all directions. In extreme cases (e.g. male orang-utans) they send pouches deep to the clavicles as far as the axillae and over the pectoral region. Each sac meets its fellow in the mid-ventral line beneath the platysma, where a septum or its remnants indicates the dual origin. The sacs are largest in orangs, but almost as complicated in gorillas. In man these outgrowths are represented by small vestiges of variable size (laryngeal saccules) passing upwards between the mucosa and the thyroid cartilage and thyrohyoid membrane. A similar situation occurs in the gibbons of the genus *Hylobates,* but in the siamang (*Symphalangus*) the sacs escape from the larynx as in the great apes and meet in the midline of the neck to unite in a globular inflatable receptacle.

Distinct saccules resembling the human structures also appear in many Old World and New World monkeys and in some (e.g. baboons and *Macaca irus*) (Bartels, 1905) they are bifid, having superior and inferior diverticula without, however, escaping from the confines of the larynx.

Regarding the functions of accessory laryngeal air-sacs, there can be no doubt that they serve to reinforce the voice and give it carrying power, but they also serve a respiratory function by acting as reserves of air to be drawn upon during bouts of excessive activity (Negus, 1949).

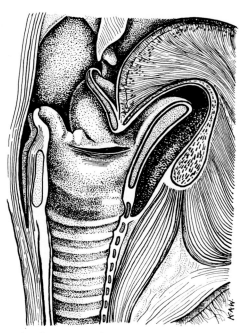

FIG. 57. Median sagittal section through the larynx of a Rhesus monkey (*Macaca mulatta*) showing the disposition of the ventral (subhyoid) median air-sac.

Lower Respiratory Tract

As the functions of the lower tract are similar in all land mammals there is, not unexpectedly, a general similarity in structure and topography. Consequently, variations occur only in minor matters such as lung lobation and patterns of bronchial ramifications (Narath, 1901; von Hayek, 1960).

The trachea, as in other mammals, is kept patent by the support of a series of C-shaped or U-shaped cartilaginous rings. In Lemuroidea, however, but not in Lorisoidea, the rings are complete and often calcified (Straus, 1931). They are also complete in *Lagothrix* (Göppert, 1931). In *Lemur variegatus* the thoracic part of the trachea is unduly long on account of the backward position of the pericardium.

The general trend in passing from lowlier to advanced Primates is for the reduction in the number of lobes in both lungs, more especially the left, and a resultant simplification in the bronchial pattern.

Most commonly the left lung presents two lobes and the right four. On the left the two lobes, apical and basal, are separated by a deep fissure, lined with pleura, that completely separates them to the hilum. The basal lobe is sometimes (e.g. in *Macaca sinica* and occasionally in man) marked by an incomplete fissure which demarcates a cardiac lappet on the ventral border. The right lung, typically four lobed, presents apical, middle, basal and azygos (or infracardiac) lobes. The fissures are not so deep as on the left, except for that between the apical and middle lobes. It is, however, often incomplete in *Macaca* and man. The middle lobe is very variable in size. The basal lobe is sometimes further sub-divided by supernumerary notches or fissures (e.g. in *Macaca sinica* and in man).

The azygos lobe is a small pyramidal process springing from the medial side of the basal lobe near the hilum. It projects to the left in relation to the thoracic part of the postcaval vein, between the pericardium and the diaphragm. It is lacking altogether in man and in the gorilla and chimpanzee, though it has been encountered on rare occasions in both apes (Sonntag, 1924; Washburn, 1950). The orang-utan is unique in the complete absence of lung lobation on both sides. In gibbons the azygos lobe is present (Sonntag, 1924). The only other primate in which its absence has been reported is *Lemur variegatus*.

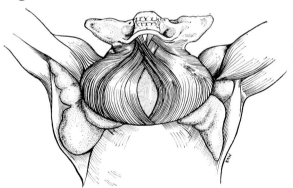

FIG. 58. Inflated air-sacs of a large orang-utan (*Pongo pygmaeus*) showing the extent of the sacs, extending to beyond the armpits and partly covered by the platysma myoides decussating in the median line. Note also the massive cheek pads.

14

Anatomy and Physiology of Reproduction

Anatomy of the Male Organs

It is very remarkable, considering that the organs have the same rather limited functions to perform, how varied the male genitalia of primates are in their morphology – a circumstance which lends support to their value in assessing relationships. These remarks apply especially to the size and location of the testes and the morphology of the penis.

TESTES

In no species of primates are the testes retained in their embryonic abdominal position, but not in all do they find a permanent position in the scrotum. In spite of contradictory statements by earlier observers (e.g. Klaatsch (1890, 1892a), Bolk (1907), Mijsberg (1923) and Weber (1928)) it has been established that a scrotal anlage is developed during late foetal life and persists thereafter, though it may be inconspicuous in early postnatal life, especially in Old World monkeys, where the testes do not descend until puberty (4–5 years of age) (Wislocki, 1936). Time of descent is variable, being earlier in man and the great apes, where complete descent occurs just before or just after birth. In *Ateles* also descent occurs in infancy. Testes in Old World monkeys remain mobile, being retractable as far as the external abdominal ring in certain emotional states (fear, anger etc.) by virtue of a powerful cremaster muscle.

Regarding the topography of the scrotum, this may be prepenial, parapenial or postpenial. The prepenial location is the most primitive, being characteristic of marsupials. Among primates a pre-penial scrotum occurs only in Tupaiidae and in some gibbons, but this position is usually only temporary to the more permanent para-penial site in the tree-shrews (Wood Jones, 1917; Hill, 1958). The parapenial position also occurs in *Perodicticus* and in marmosets and tamarins and is regarded as a transitional state towards the more usual post-penial location of the sacs in the majority of Primates.

The scrotum is not invariably as pendulous as in man, the only comparable case being *Alouatta* (Wislocki, 1936). In most prosi-mians, monkeys and apes it is a globular or oval sessile sac not, or only a little, constricted at its neck.

The size of the testes and hence of the scrotum is very variable, differing even among related forms. For example they are relatively huge in the chimpanzee, ridiculously small in the gorilla. In maca-ques they tend to be large, though variable in some (e.g. the Rhesus monkey) according to season (Sade, 1964). The relative size of the testes in comparison with body weight has been recorded by Schultz (1938b) in a large series of simians. In New World species he found the relative testicular weight fluctuated between 0·1 (*Aotes*) and 0·4 (*Alouatta*). Among Old World forms the weight is relatively greater among the Cercopithecidae than the Colobidae. The testes are small, but undergo seasonal enlargement in *Loris*. In *Galago, Arctocebus* and *Perodicticus* the scrotum is thick walled and glandular. In some Lemuroidea (e.g. *Lemur, Hapalemur*) the scrotum is hairy; in others (e.g. *L. catta*) naked.

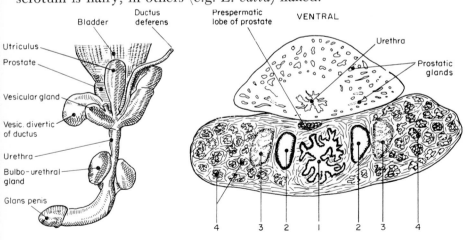

FIG. 59. *Ptilocercus lowii*, internal reproductive organs of the male Pen-tailed Tree-shrew ×2·5, from Eckstein. On the right the prostatic urethra is shown (aprox. ×14) in transverse section. 1. Utriculus prostaticus; 2. Ductus deferens; 3. Diverticulum of ductus; 4. Vesicular gland.

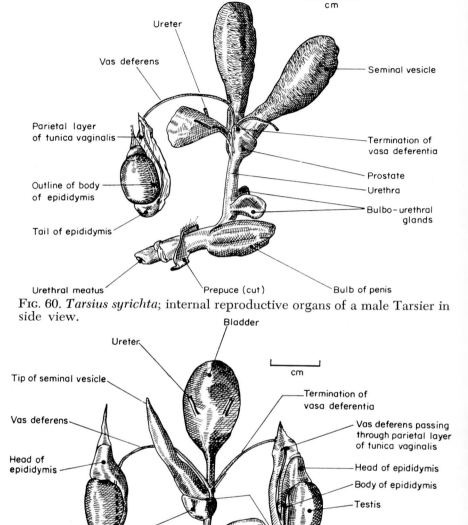

FIG. 60. *Tarsius syrichta*; internal reproductive organs of a male Tarsier in side view.

FIG. 61. *Loris tardigradus*; internal reproductive organs of the male Slow Loris, drawn from the dorsal aspect. The penis shown turned to the side.

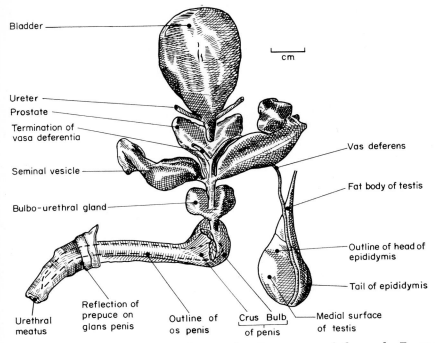

Bladder

cm

Ureter
Prostate
Termination of
vasa deferentia

Vas deferens

Seminal vesicle

Fat body of testis

Bulbo-urethral gland

Outline of head of
epididymis

Tail of epididymis

Urethral
meatus

Reflection of
prepuce on
glans penis

Outline of
os penis

Crus Bulb
of penis

Medial surface
of testis

Fɪɢ. 62. *Lemur fulvus*; internal reproductive organs of the male Brown lemur, drawn from the dorsal aspect. The penis is shown turned to the side.

Tʜᴇ Pᴇɴɪs

To catalogue all the penile variations met with among Primates would need a book almost as large as this one, so we must be content with generalities and some characteristic examples. For details the works of Meisenheimer (1921), Ottow (1955) and Hill (1958) and references there cited should be consulted.

The intromittent organ is generally of cylindroidal form, but varies in regard to the differentiation of a terminal glans. The distal part is enclosed in a cutaneous sheath (prepuce) which may or may not leave the glans exposed. This part is invariably pendulous in the flaccid phase, a feature which the Primates share only with bats and bears. Furthermore, in all Primates except *Tarsius, Lagothrix* and man, the distal part of the penis is strengthened by a rod-like bone (*baculum, os penis, os priapi*) which varies in form, being incipiently or frankly bifurcated at the tip in many prosimians.

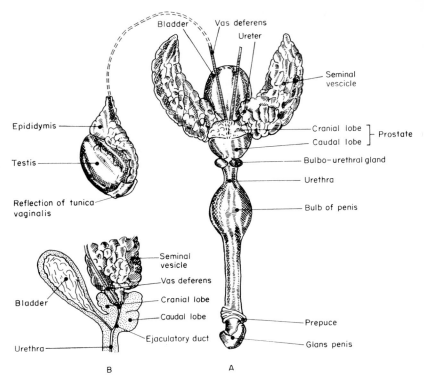

FIG. 63. *Macaca mulatta*; A. Male internal genital organs from the dorsal aspect indicating the general pattern of the organs in any simian. B. Enlarged view of the prostatic urethra in sagittal section.

A further specialization found in many prosimians is the surface adornment of horny spicules or recurved spines that act as grappling spurs (e.g. in *Galago*, *Arctocebus*, *Lemur* and the Indriidae) disposed in varying ways and subject to minor intergeneric variations in size and complexity. In *Loris*, *Nycticebus*, *Perodicticus* and *Cheirogaleus*, the spines are represented by a polygonal pattern of minute keratinized papillae. The same applies to *Tarsius* where in other respects the organ shows a curious resemblance to the human counterpart (Wood Jones, 1929), though it bears a conical apex and some neighbouring modifications (Hill, 1958; Le Gros Clark, 1959).

In simians the penis is comparatively simple in form. In the New World species the glans is not well differentiated, the tip of the penis presenting in marmosets a slight rounded or oval enlargement, but in Cebidae it is truncated and smooth, except in *Ateles* where numerous keratinized barbs are present directed away from the flattened or slightly excavated apex.

In Old World monkeys the glans is well differentiated and marked off from the shaft by a deep retrocoronal sulcus. The glans is typically acorn-shaped like the human structure, but tends to extend proximally on the dorsum to a greater extent than on the perineal side. Laterally the corona is generally marked by notches. Sometimes a median dorsal notch is also present. There are frequently minute horny papillae or spicules on the shaft of the penis or on the glans, or even both. Gross modifications in the morphology of the glans distinguish some of the species of *Macaca*. An expanded glans is lacking from the penis of orangs and chimpanzees; in the former the organ ends in a bluntly rounded apex, whereas the chimpanzee penis is long, slender and almost pointed. The gorilla, on the contrary, has a small penis with an ovate glans. All the Old World monkeys and apes have a baculum.

Genital Tract

The only parts of the internal male genitalia needing comment are the accessory glands. Seminal vesicles vary in size according to season and are usually large in *Tarsius, Macaca* and baboons (but not in geladas). They are entirely lacking in *Daubentonia*. The prostate is generally bilobed, a transverse groove dividing it into a cranial and a caudal lobe, but only in man and *Ptilocercus* does it completely encircle the urethra. In other primates it leaves the ventral wall of the passage uncovered (for details, consult Mijsberg (1923)).

The mode of termination of the genital ducts shows some interesting variations. In Lorisoids the ductus deferens and vesicular ducts open separately into the urethra – except in *Galago,* where they join to form a common ejaculatory duct as they do also in *Lemur.* In *Hapalemur,* however, they remain distinct but share a common opening on reaching the urethra. In the Indriidae they are separate, as in *Loris,* and the terminal part of the vas deferens is thickened and glandular. A common ejaculatory duct is found in *Tarsius* and in monkeys and apes. In *Tarsius* a small, blind diverticulum is found between the ejaculatory ducts and opening separately from them into the urethra.

Anatomy of the Female Organs

OVARIES

These are more or less elongated oval smooth surfaced organs lying on the side wall of the pelvis in adult Primates. They vary in size, being relatively larger in New World monkeys than in other forms (Wislocki, 1932). They also undergo cyclical enlargement, at any rate locally with the regular ripening of the egg-follicles. Rupture of ripe follicles with the escape of the contained ovum is followed, as in mammals generally, by the formation of a corpus luteum. In platyrrhines and gibbons luteal tissue occurs interstitially. If pregnancy supervenes, the latter persists and serves to maintain, by its internal secretion, the condition of the genital tract suitable for continuation of the pregnancy. Otherwise the corpus luteum soon atrophies.

GENITAL TRACT

In contrast to the male, it is in the genital passages that, in spite of basic similarity, the greatest amount of anatomical diversity is found. The usual mammalian division of the tract into Fallopian tubes, uterus and vagina is found. Fallopian tubes have the usual parts, but may be much coiled or, as in man, marmosets and *Saimiri*, relatively straight. Along its cranial wall in prosimians and in New World monkeys, a peritonial fold (superior mesosalpinx) extends laterally from the uterus and, at the ovarian end, may expand to form an ovarian bursa, partly or completely covering the ovary dorsally. This generally is not or only feebly developed in *Tarsius* and Old World monkeys and apes; there is, however, an ovarian bursa in *Procolobus* (Hill, 1952).

The uterus of the prosimians and of *Tarsius* is bicornuate, i.e. it consists of a short median body (*corpus uteri*) situated caudally and two conical horns (cornua) cranially the latter receiving the Fallopian tubes. This, of course, is an advance on the completely paired uterine segments of many lower mammals, but not as advanced as the single median uterus of the higher Primates.

The bicornuate uterus of *Tupaia* has a very short corpus and long cornua, a more primitive arrangement than in lorises, lemurs or *Tarsius* (Le Gros Clark, 1934). The uterine cavity in *Tupaia* is separated from that of the vagina by a well marked thickening (external os) (Wood Jones, 1917).

In lorisoids and lemuroids the corpus uteri is about the same length as the cornua and may exceed them (e.g. in *Lemur, Lepi-*

lemur) but in *Perodicticus, Hapalemur* and *Indri* the cornua are relatively long and pointed (Eckstein, 1958).

In *Tarsius*, the bicornuate uterus closely resembles that of *Lemur*, but the lateral horns are relatively shorter and fusiform. A cervix uteri is differentiated, so an advance on the lemuroid arrangement is evident.

In all simians the uterus consists of a single median body (without cornua) and a neck or cervix. Traces of the original bilateral origin are sometimes found as a groove incising the uterine fundus dorso–ventrally (e.g. *Macaca mulatta*) (Wislocki, 1933). In general the cervix is more developed in monkeys than in the apes and man, being often longer than the corpus and complex in structure (e.g. in *Macaca*). It projects into the vault of the vagina as a rounded knob (pars vaginalis cervicis); this is specially prominent in *Ateles* (Wislocki, 1932) and very well marked in marmosets. In *Aotes*, however, the cervix is small and lacks the pars vaginalis, a condition

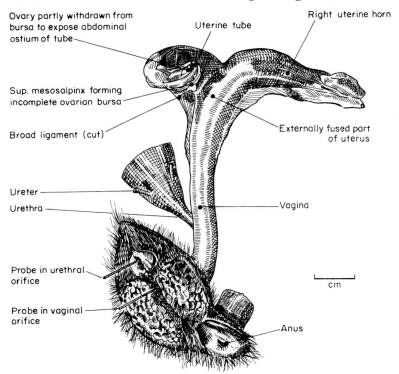

FIG. 64. *Perodicticus potto*; female reproductive tract from the dorsal side, with the caudal end of the vagina twisted over to demonstrate the external genitalia.

that reappears in *Lagothrix* and in gibbons. In *Macaca* the pars vaginalis is variable in size and position; upon it the slit-like external os lies nearer the ventral than the dorsal wall, so that the dorsal lip is thicker and the dorsal fornix deeper. In *Papio*, the reverse has been recorded (Eckstein, 1958), but in *Presbytis* the relations are as in *Macaca* (Ayer, 1948).

In the apes, the resting uterus is small in gibbons and orangs and larger in chimpanzees, where it is smaller than in man, whilst in the gorilla the organ is larger than in the human female. In

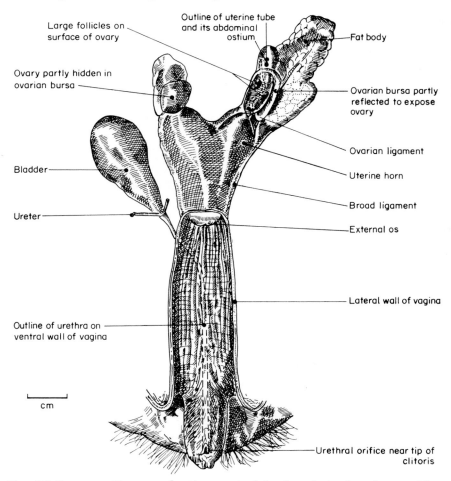

Large follicles on surface of ovary

Outline of uterine tube and its abdominal ostium

Fat body

Ovary partly hidden in ovarian bursa

Ovarian bursa partly reflected to expose ovary

Ovarian ligament

Bladder

Uterine horn

Broad ligament

Ureter

External os

Lateral wall of vagina

Outline of urethra on ventral wall of vagina

cm

Urethral orifice near tip of clitoris

FIG. 65. *Lemur catta*; reproductive tract of the female in dorsal view. The vaginal tract has been laid open from above to expose the long clitoris and the position thereon of the urethral aperture.

general shape and structure the organ in the large apes resembles that in man.

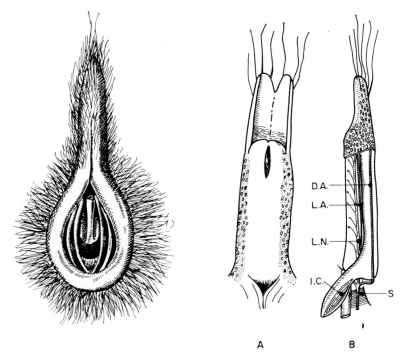

Fig. 66. Plan of the external genitalia in a female *Microcebus murinus*, Dwarf Mouse-lemur (left) and *Hapalemur griseus*, ♀ Grey Gentle lemur (right).

EXTERNAL GENITALIA

In female Primates these are built upon the same general plan, but vary considerably in details. The external opening of the female genital tract lies some distance ventral to the anus, the intervening tissues constituting a "perineal body". In the lowest Primates, as in many lower mammals, the opening leads to a vestibule, of varying length, which receives both the vaginal and urinary outlets. In the more advanced forms the vestibule becomes progressively shorter and shallower until reduced to a sagittally directed median groove or depression (*rima pudendi*) with vaginal and urethral openings a little distance apart in its floor. Ventral to the urethral opening a clitoris, or small homologue of the penis, is invariably present, but it is extremely variable in prominence and its relation to the urethra.

E

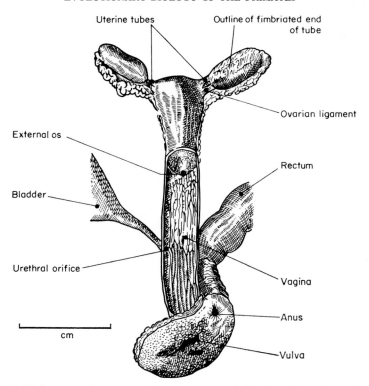

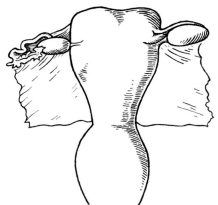

FIG. 67. *Callithrix jacchus*; reproductive tract of the female in dorsal view. The caudal end of the vagina is twisted over to expose the vulva. From Eckstein.

FIG. 68. Uterus and adjacent structures of a Moustached monkey, *Cercopithecus cephus*, as an example of a simian reproductive tract. Redrawn from Hill (1966, p. 338).

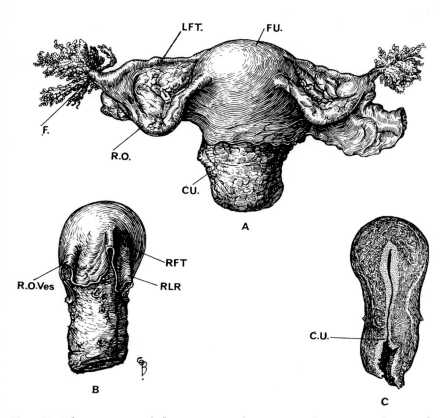

Fig. 69. Three views of the non-gravid uterus and its appendages of a Gorilla (*Gorilla gorilla*). In A. the appendages are shown complete in ventral view, with the ovaries and Fallopian tubes. In B. the uterus is shown alone in right lateral view and in C. the organ is indicated in sagittal section. Redrawn from Wislocki.

CU.	cervix uteri	RFT.	right Fallopian tube
F.	fimbriated end of tube	RLR.	right ligamentum teres
FU.	fundus uteri	RO.	right ovary
LFT.	left Fallopian tube	ROVes.	right ovarian vessels

Surrounding, or partly marginal to, the rima and its contents there are, usually developed to some degree, lateral cutaneous elevations or labia. Immediately bordering the rima are thin, hairless folds (labia minora) which ventrally come into relation with the clitoris, enclosing it in a sheath or hood (praeputium clitoridis). Peripheral to these occur, in many cases, thicker folds (labia majora) of more normal, even hairy skin, that tend to unite ventrally around the root of the clitoris in a commissura labiorum (for consideration of the confused homologies with respect to the labia, see Hill (1958)).

In *Ptilocercus* labia minora are developed along both sides of the rima and also form a broad commissure along its ventral extremity. A small clitoris is joined to their inner aspect by thin folds of mucous membrane. Lateral to these labia are ill-defined elevations of hairy skin corresponding to labia majora, but these fade out about half way along the length of the rima, being better developed dorsally than ventrally.

Tupaia, in contrast, has an elongated pendulous, penis-like clitoris covered with sparsely haired skin. Its perineal surface is marked throughout by a groove that widens dorsally to form the rima containing a large vestibular opening into the vault of which the vagina is continuous, the urethral orifice being on the ventral vestibular wall. There is no evidence of labia. A small vesicle was found by Pehrson (1914) near the tip of the clitoris in *T. javanica*.

A remarkable feature in the lorisoid prosimians is the great enlargement of the clitoris and the fact that the urethra tunnels through it to open at its tip. In *Galago* it is provided, as in the male, with a baculum. The same state of affairs has been developed in *Hapalemur*, where a baculum is also present, but in all other lemuroid prosimians the urethra opens within the rima at the base of the clitoris.

Microcebus is rather similar to *Tupaia*, but swollen labia minora encircle the rima from the base of the clitoris. In *Lemur* too there is a long pendulous clitoris, but the picture is complicated by the cutaneous, glandular foldings on the perineal body (except in *L. catta*). A small baculum has been recorded in the organ of *L. variegatus* (Pohl, 1910). In the Indriidae, the clitoris is elongated and narrow in *Avahi*, shorter and stouter in *Propithecus*, expanded and leaf-shaped in *Indri*. The glans clitoridis is incipiently bifid in *Daubentonia* and *Tarsius;* in the latter there are well marked labia minora which largely conceal the short clitoris.

Marmosets and *Callicebus* are provided with a short, inconspicuous clitoris and in the former the labia minora are bulky from glandular hypertrophy, typical labia majora being absent. In other New World genera, such as *Cebus, Saimiri, Lagothrix* and *Ateles*, the clitoris is large and pendulous rendering sex determination difficult, especially as the rima is short, inconspicuous and without labia or only small ones.

In Old World monkeys the clitoris varies in prominence, but is never so elongated as to become pendulous. The genitalia in *Cerco-*

pithecus and its allies are inconspicuous, except in the small Tala-poin (*Miopithecus*). In mangabeys, the macaques and baboons, the clitoris is short, but provided with a rounded prominent glans of bright red tint. Labia are variably developed, being better defined in infants and juveniles than in sexually mature females. A promi-nent feature in the genital region of many female Old World mon-keys is the periodic development of an area of swollen (oedematous) skin (sexual skin) involving the labia, the preputium clitoridis, the circumanal tissues and, in some cases, more peripheral skin in-cluding that around the callosities and the root of the tail and even parts of the lower back and thighs. The swelling appears slowly during the period when ova are ripening in the ovary but, after ovulation has occurred and the luteal hormone is released, the swelling rapidly subsides. This phenomenon occurs in *Miopithe-cus*, mangabeys, all baboons and many macaques, but is not known in *Cercopithecus* or the Asiatic langurs. In some African cobobines however, a similar swelling affects both sexes and once developed appears to persist, so it is evidently of a different nature from that in baboons. The form taken by the sexual skin swelling differs among the species of a single genus, e.g. *Macaca*, where it is very limited in *M. irus* and then only in relatively young females; absent altogether in *M. sinica* and *M. arctoides*; more extensive in the Rhesus (*M. mulatta*) where it is, however, individually variable and reaches grotesque proportions in *M. nemestrina*, in which respect it resembles the baboons of the genus *Papio*. On the contrary, the swelling is more limited and circumscribed in *Mandrillus* and in *Cercocebus* (for details consult Hill (1958) or Eckstein and Zucker-man (1962)).

In anthropoid apes, labia minora are generally present, but labia majora occur only in foetal and juvenile stages, thereafter under-going regression, so that they are absent or inconspicuous in adults. However, they are often encountered in orangs. Ontogenetic reduc-tion proceeds furthest in gorillas and least in gibbons. Pocock (1926) found labia minora well developed in chimpanzees and some gib-bons. A true sexual skin undergoing cyclical enlargement occurs only in the chimpanzee, where it affects the perineum and circum-anal tissues.

Physiology of Reproduction

Reproductive mechanisms are as varied as the anatomical features of the organs involved and hence serve equally well in assessing relationships.

It is a well established fact that among mammals, other than man, reproductive activity is restricted to certain periods of the year referred to as the "breeding season" – a term liable to confusion, since some interpret it as the season during which the young are born, others use it to indicate the period when the sexes come together. These may be many months apart depending upon the duration of pregnancy (gestation period).

Breeding, i.e. the giving birth to young, in any one locality is closely correlated with the local climatic conditions insofar as young tend to be produced at a season most advantageous to their welfare and growth. This still applies where, through the agency of man, a species is transported from one country to another or from one hemisphere to the other across the equator, inducing a complete reversal in the breeding season (e.g. in *Lemur catta*). Individual animals, however, may keep to their internal sexual rhythm for a year or two until the change takes place – probably through a change in the effect of the duration of daylight on the internal secreting glands which control the activity of the sex organs. Since most Primates are inhabitants of the tropical belt, where there is little seasonal climatic change apart from rainfall, it is not surprising to find that young are born during all months of the year, though there may be seasonal peaks in the numbers produced. It is only in species which inhabit northern forests (e.g. Japan) or high altitudes – as in the Himalayas – that restricted breeding seasons are in operation.

In mammals with markedly restricted breeding seasons, the gonads (ovaries or testes) during the greater part of the year are small and the genital passages shrunken and inactive. This condition is described as *anoestrus*. At the approach of the sexually active season the ovaries enlarge and some of their unripe ova mature and the follicles containing them enlarge and produce an internal secretion (*follicular hormone, folliculin* or *oestrin*). The effect of this is widespread, including changes in the genital tract and also influencing the animals' behaviour. This phase constitutes *prooestrus* or period when the female is "coming into heat". When ovulation takes place – usually spontaneously – and the follicular hormone ceases to act, the hypertrophied mucous lining of the genital tract undergoes degeneration. This is the phase known as *oestrus* or "heat"; this is the only time when the female will receive the male – whose testes at the same season have become enlarged and are actively producing spermatozoa and when he is said to be in *rut*. Meanwhile the corpus luteum has replaced the ovarian follicle; this produces a hormone (*lutein* or *progesterone*) which promotes further hypertrophy of the lining of the genital passages and stimulates maternal behaviour and responses for, in the natural course of events, pregnancy supervenes and thenceforth cyclical activity ceases until after the pregnancy has terminated. If the female fails to become pregnant a short phase of so-called pseudo-pregnancy (or *metoestrus*) is manifested, but soon the corpus luteum degenerates and its effects on the mucosa demolished. Then follows a phase of regeneration and the female returns to the stage of anoestrus – until the following year in most mammals. In others, however, (e.g. in cats) the cycle recurs twice a year; such are said to be dioestrous. Still others (e.g. some rodents) are polyo-estrous; in these after a short anoestrus, ovarian activity recom-mences and prooestrus follows, so that a succession of oestrous cycles succeed each other throughout the year.

The problem now arises as to how events in the reproductive life of Primates (including man) may be homologized with the above pattern and into which oestrous category the different species fall. In other terms the question resolves itself into how the men-strual cycle in Primates can be related to the oestrous phenomena of lower mammals.

This question is answered by the recognition of menstrual bleed-ing as an exaggeration of the degenerative phenomena occurring in the mucous lining of the uterus (endometrium) at the inception of the luteal phase of the ovarian cycle. It is exaggerated insofar as the endometrial loss is of such magnitude as to erode into the blood vessels, enabling blood cells to escape from the passages along with other debris. Menstruation therefore corresponds to metoestrus of a lower mammal. Its manifestations, however, vary greatly over the primate series, as will be seen hereafter.

In *Tupaia* there is evidence of a breeding season extending over at least 8 months (June–January) (Zuckerman, 1932b). Copulations occur throughout the year, but data on births suggest seasonal peaks, births being maximal during summer months when sexual cycles are minimal. Oestrous cycles of 10–12 days and 20–22 days have been observed and menstrual bleeding has been reported at the end of a pseudo-pregnancy, 2–4 days subsequent to a post-pseudo-pregnancy oestrus (progesterone-withdrawal bleeding) (Con-away and Sorenson, 1965; Sorenson and Conaway, 1964, 1968). Gestation in *T. longipes* lasts 47 days.

The lorisoids are generally believed to have a sharply defined breeding season recurring every 6 months. This certainly applies in *Loris*, where in the male the rut is associated with enlargement and descent of the testes, combined with changes in the scrotum (personal observations in Ceylon). From studies of the oestrous cycle of females in Europe, Manley (1966) concluded that seasonal delimitation of breeding is less rigid than commonly supposed. *Nycticebus,* on the contrary, has no restricted breeding season. Females come into season in Malaya every 2 months and births have been recorded in all months. The potto, likewise, has been reported as cycling continuously throughout 20 months of observation (Ioannou, 1966; Manley, 1966). *Arctocebus* is similar, with births occurring at all seasons, though more occur in the middle and end of the dry season and start of the wet season than at other times (Jewell and Oates, 1969). The gestation period is 131 days (Manley). Butler (1957) concluded that, in the Sudan, *Galago senegalensis,* has a restricted breeding season, thus confirming earlier and some later reports (Lowther, 1940; Herlant, 1961; Petter-Rousseau, 1962; Ansell, 1963). In 1960, however, Butler, with wider facilities, revised his opinion admitting that oestrous phenomena recur throughout the year and that estimates of restricted breeding are derived from seasonal peaks in the birth rate. Nevertheless in 1964, Butler inclined to the view that there are two yearly breeding seasons at 6 month intervals, later (1967) confirming this for the wild populations in the Nuba mountains and in northern Uganda. *Galago senegalensis moholi,* in the southern hemisphere, comes into oestrus twice yearly, between March and August, the second oestrus being an immediate post-partum oestrus. Twins are produced twice a year (between July and January) after a gestation of 122–125 days (Doyle *et al.,* 1967). Some evidence of lunar periodicity has been noted in some lorisoids as well as lemurs (Cowgill *et al.,* 1962).

The Madagascar lemurs are strictly seasonal in their sexual behaviour, though many details remain obscure. Latest data, based on field studies, are from Petter (1965). Sexual interest is accompanied by increase in scent marking activity and by pairs sleeping huddled together. Much mutual licking also occurs during sexually active phases, the females especially attempting to excite the males. Females will approach a sleeping male, licking him or standing motionless with stiff legs and her tail arched over his. Mating probably occurs at dawn; in captivity *Cheirogaleus major* has been observed in copulation, the male holding the female's waist and briskly licking her neck and flanks; he grips the female's ankles with his feet, emitting a twitter and undulating his tail.

The mating season in *Microcebus* is from September to January; the gestation period being 2 months, with a litter of 2 or 3. *Lemur* species mate between April and June and produce a single offspring

4½ months later. *Lepilemur* is similar except that mating occurs later (May–July). *Propithecus* has the mating period between January and March, and produces a single young after 5 months gestation. According to native testimony, the aye-aye (*Daubentonia*) produces a single young between early February and late March, the gestation period being unknown. Captive births of *Lemur* in London almost all occurred between March and June (Zuckerman, 1932b) and similar results were observed at Giza (Egypt) (Flower, 1933), thus indicating a reversal of the breeding season on crossing the equator.

Tarsius, on the contrary, appears to be capable of breeding at all seasons, since pregnant uteri have been collected in the field during all months (van Herwerden, 1905, 1925; Zuckerman, 1931, 1932b). Oestrus occurs cyclically and is accompanied by congestion and turgidity of labia minora and keratinization of the vaginal mucosa lasting 24 hours. This phase is followed by deturgescence of labia and replacement of cornified cells in the vaginal washings by leucocytes (Catchpole and Fulton, 1943; Hill *et al.*, 1952). The full cycle lasts from between 23 and 28 days.

Marmosets, again, are capable of breeding at all seasons, though peaks in births occur in spring and autumn, after a gestation of 4½ months. Twins are normally the result.

Other New World monkeys (Cebidae) have been said to show restricted breeding seasons. This appears true in *Saimiri oerstedi*, but the reverse is the case in *Cebus*, *Alouatta* and *Ateles* (Wislocki, 1930). Ovulatory cycles occur in *Cebus* throughout at least 9 months of the year; most births, however, are confined to two periods, May–June and October–November (Hamlett, 1939). Menstrual bleeding has commonly been denied for this family and they certainly differ from Old World forms in this respect, but slight cyclical bleeding lasting 2–3 days, recurring at intervals of 3, 6 or 10 weeks has been noted in *Cebus* since the time of Rengger (1830) and confirmed by Hamlett. Histological evidence of cyclical changes in the endometrium, in the absence of observable or scanty bleeding, has been brought forward for *Cebus*, *Alouatta* and *Ateles* (Howard, 1930; Dempsey, 1939; Kaiser, 1947). Gestation in *Cebus* has a duration of 6 months and in *Ateles* averages 139 days (Kenneth 1947). Sexual activity is suppressed in lactating females.

In Old World monkeys menstrual cycles are regularly found, but there are many differences in detail. In *Cercopithecus*, of which only a few species have been studied, differences exist between species, while the rhythm is influenced by captivity and other environmental factors (Rowell, 1970). Butler (1966a) reported an obvious menstrual flow lasting 2–3 days at irregular intervals (24–149 days) in *C. aethiops* (grivet) without any sexual skin changes. Mid-cycle bleeding was observed once. Rowell (1970) studied another form of *C. aethiops* (vervet) in Uganda and found external bleeding unusual, but recognizable by vaginal lavage. Intervals varied from 25 to 46 days, but scanty mid-cycle bleeding sometimes occurred. Births occurred especially between January and June, but some babies were produced in July and October, showing no relation to wet and dry seasons. On the other hand, at Amboseli (Kenya) Struhsaker (1967) found the birth season of *C. aethiops* to be between October and March (Zuckerman, 1932a).

Externally visible menstrual bleeding is sometimes encountered in *C. mitis*, but in most cases it is scanty even on lavage. In females of the subspecies *C.m. erythrarchus* and *C.m. kolbi* it was copious and regular, but unaccompanied by other external signs, beyond some turgidity of the labia. (personal observations). Rowell (1970) found menstrual manifestations recurring at average intervals of 30 days. Pregnancy duration is around 140 days and births, in contrast to *C. aethiops*, are restricted to 2 periods, June–July and November, both outside the rainy seasons. On the other hand, in the wild *C. mitis* in the Aberdare Mountains of Kenya a single breeding season with conceptions occurring during the "small rains" (August–January) and births just before the "heavy rains" (January–April).

West African species of *Cercopithecus* (*C. nictitans, C. pogonias, C. cephus* and *C. neglectus* appear to concentrate their births in the period December–April i.e. the short hot dry season. (Gautier-Hion, 1968; Sabater Pi, 1970).

Erythrocebus appears to resemble *Cercopithecus* in its reproductive phenomena, but *Miopithecus* females have their sexual cycles marked by a voluminous sexual skin, as well as macroscopic menstrual bleeding. Although there is no strict correlation between bleeding and the phase of sexual swelling (Tomilin, 1940). In West Africa they bring forth their young in the hot, dry season, as with *Cercopithecus*. *Allenopithecus* agrees with the talapoin in exhibiting a cyclical sexual swelling; births, in captivity, have occurred in June, July and September (Hill, 1966).

The mangabeys (*Cercocebus*) have a mean cycle length of approximately 30 days. Menstrual bleeding is copious and a characteristic lobate sexual swelling takes place. The gestation period is said to be 213 days (Jennison, 1927).

Macaques vary as regards the occurrence of a sexual swelling (see above p. 123) but all adult females undergo regular menstrual cycles over a period (in the Rhesus, *M. mulatta*) of 15 years during which ovulations take place. Females of *M. mulatta* also exhibit periodic reddening of the hind quarters and of the face; during pregnancy these cutaneous phenomena persist. The redness usually attains maximum intensity at or soon after ovulation, i.e. during the 3rd week after onset of menstruation, fading rapidly before the next flow unless pregnancy supervenes. The gestation period ranges between 149 and 180 days (Hartman, 1932) or an average of 168 days (van Wagenen, 1945). Based on captive specimens, the conclusion has been generally held that Rhesus monkeys are capable of breeding at all seasons of the year (Corner, 1923; Zuckerman, 1930), but some observations on natural populations (Heape 1896; Hingston, 1920) or on recently captured material (Hartman, 1931) suggest that the birth season is restricted to September–October. Moreover, populations inhabiting higher altitudes in North India, where seasonal climatic changes are more rigid, also appear to show restricted breeding seasons.

In wild populations of Rhesus monkeys copulations have been observed in all months except March, with greatest frequency of sexual behaviour between October and December; births and infants most frequently observed from March to June, with a few in September. From November to March no newborns were recorded (Southwick *et al.*, 1961).

Seasonal variations in testicular size have been observed in male *M. mulatta* (Sade, 1964).

The Japanese macaque (*M. fuscata*) shows a restricted breeding season, copulations occurring between October and April, with majority of matings in January and February. Births occur in May–September, with only a few in April and October and none between mid-October and mid-April. (Mizuhara, 1957; Kawai, 1962; Lancaster and Lee, 1965). Menstrual cycles, accompanied by external bleeding, occur throughout the year but oestrus, characterized by behavioural features, is limited to the mating season (Tokuda, 1962). Swelling of the sex skin, though claimed to occur by Anderson (1878) was not encountered by Pocock (1926). According to Tokuda some puffiness accompanies oestrus, especially in younger females, but reddening of the perineal skin and of the face occur as in *M. mulatta*.

Other species of *Macaca* exhibit similar menstrual phenomena to the species above discussed, but differ among themselves in the presence or absence of cyclical swelling of the sexual skin (as mentioned on p. 123). Definitive swelling, as distinct from mere temporary puffiness, occurs in *M. sylvanus, M. silenus, M. nemes-*

trina, M. maurus and *M. cyclopis*, but is lacking in *M. irus, M. arctoides, M. radiata* and *M. sinica.* In the last two mentioned a specialization of the cervix uteri results in the production of a large amount of odorous mucus secretion that escapes from the genital passages at all times though varying periodically in quality and quantity. In the Pig-tailed macaque (*M. nemestrina*) sexual swelling reaches grotesque proportions resembling thus baboons of the genus *Papio.* Sexual maturity is attained at 50 months. Menstrual cycles recur at 32–40 days and the average gestation period is 170 days (Kuehn *et al.,* 1965). The Celebesian Black Ape (*Cynopithecus niger*) exhibits the same external manifestations as *M. maurus,* but with some minor variations in the form of the sexual swelling.

Baboons of the genus *Papio,* have been studied in detail both in the field and in captivity (see especially Eckstein and Zuckerman (1962) and works there cited, also Saayman (1970)). Menstrual cycles recur every 33–39 days, the bleeding having a duration of 3–4 days. A large sexual skin swelling commences to develop during or just after the menstrual flow, reaching its peak in one week, showing minor differences in the 5 species concerned. Copulations take place in all months, but tend to be limited to females in the swollen condition. Births also occur throughout the year, but may show peaks in some species (e.g. *P. hamadryas*). The gestation period averages $172 \cdot 2$ days in *P. hamadryas,* 187 days in *P. ursinus.*

Baboons of the genus *Mandrillus* differ from *Papio* in the morphology of the sexual swelling which, on the contrary, resembles that seen in Mangabeys (Hill, 1970). Regular menstruation at approximately 6 weekly intervals was observed in *M. leucophaeus* by Percy (1844), and more recently by Zuckerman (1937) who reported a mean cycle length of $36 \cdot 2 \pm 0 \cdot 9$ days.

Geladas (*Theropithecus*) are peculiar in dispensing with the usual sexual skin phenomena. Instead the female develops a series of vesicles around the periphery of the perineum and also around the characteristic naked area of skin on the throat and chest. During the premenstrual (follicular) phase of the cycle, the bare skin areas become intensely reddened and the pearly vesicles enlarged. At the approach of menstruation the skin diminishes in colour to pink and tumefaction is reduced, but the vesicles enlarge and fill with watery fluid. This passes off in a few days and after vesicular recession menstruation is established, but is slight compared with *Papio,* the flow persisting for 2–3 days. The full cycle has a duration of 32–36 days. Gestation period is approximately 6 months (see Hill (1970) and references there cited).

Among the Colobidae, females of *Semnopithecus* and *Kasi* under-go menstrual cycles at approximately monthly intervals; the men-strual flow is slight, lasting 2–4 days and there are no external changes in the perineal area (Heape, 1894; Hill, 1936). In those parts of India where an annual cold season occurs, leaf monkeys of the subgenus *Trachypithecus* are said by McCann (1933) to experi-ence an annual breeding season, most infants being born in Feb-ruary and March subsequent to copulations in September and Oc-tober. In *Colobus* breeding is believed to continue throughout the year, but the gestation period had not been recorded (Napier and Napier, 1967).

Gibbons, in the wild, are stated by Carpenter (1940, 1941) to experience a restricted reproductive period, while in captivity the females often exhibit long periods of amenorrhoea. They attain reproductive maturity at 7–8 years when their cycles average 28 days duration, with a menstrual flow lasting 2 days or slightly more. No external swelling occurs, but Carpenter has mentioned some colour changes and tumescence of the labia (see Eckstein and Zuckerman (1962) and references there cited).

Among the great apes information is most detailed for the chim-panzee. Females experience regular cycles of around 36 days dura-tion, but varying individually and even at different times in the same female. A prominent sexual swelling is a conspicuous feature at one stage in the cycle and its variations make possible the sub-division of the cycle into 4 unequal phases:

1. A stage lasting one week; "pre-swelling phase" corresponding to the follicular phase.
2. Stage of swelling lasting 18 days, characterized by genital tumescence and continued follicular activity.
3. Post-swelling stage, a period of detumescence of 10 days dura-tion occupying the luteal phase of the ovarian cycle.
4. Menstruation – duration 4 days.

Average age of first menstruation is 9 years; cycles continue for upwards of 20 years. The gestation period is 250 days. Lactation may continue for over 2 years (for further details see especially Yerkes and Yerkes (1929) and Eckstein and Zuckerman (1962) and works there cited).

The Orang-utan differs in lacking external changes during the cycle, but labial tumescence develops in pregnant females shortly before they are due to deliver (Schultz, 1938a). It shrivels after parturition.

The cycle of the gorilla is inadequately known, for external changes are slight. Puberty occurs at 9 years and the cycles vary in length, the mean for 6 cycles of a captive female being 43 days (Noback, 1936, 1939). Raven (1936) observed a wild female with a pale tumescence resembling in size that of a chimpanzee, but whether this was a pregnant female or not was not ascertained.

Foetal Membranes and Placenta

In broad outlines, there is to be found among Primates a progressive improvement in the elaboration of foetal envelopes and especially in the intimacy between foetal and maternal tissues, from the lower to the higher Primates which thus bear out the concept of their representing an evolutionary series (see Chapter 2, p. 19). There are, however, some complexities and problems, especially insofar as the Tupaiidae appear to show a more intimate relationship between foetal membranes and endometrium than the lemurs and lorises.

Grosser (1933) classified eutherian placentation on the basis of degree of intimacy between the outermost foetal envelope (the chorion) and the tissues composing the endometrium. Two of Grosser's four categories are found among Primates :

    (i) the epithelio-chorial and
    (ii) the haemochorial.

In the former the chorionic epithelium, by virtue of its outer trophoblastic layer, invades the whole of the endometrium to a limited depth so that the placenta is diffuse rather than limited to one or two areas of the chorion. Maternal and foetal blood-streams are separated by the respective vascular endothelia and the surface epithelia of chorion and endometrium – an appreciable barrier to the passage of nutrients and oxygen exchange. After birth has taken place the foetal tissues separate from the maternal at the line between the chorionic and uterine epithelium without any loss of maternal tissue and no loss of maternal blood. This type of placenta is referred to as non-deciduate.

In the haemochorial type of placentation the invasive action of the trophoblast extends deeper and its intimate relation to the maternal tissues become limited to one or two oval areas whereon the eroding action of the chorionic villi penetrates through the walls of the endometrial vessels enabling them to be bathed in maternal blood, so that fewer layers separate maternal blood from foetal blood. After birth some maternal tissue and blood is lost and such a placenta is said to be deciduate and the endometrium referred to as decidua.

The foetal membranes of *Tupaia* are known from the researches of Hübrecht (1899) later supplemented by the observations of Meister and Davis (1956, 1958), Hill (1965), Verma (1965) and Luckett (1968). Arrangements are unique among mammals and fundamentally different from the conditions in all strepsirrhines. A placenta is developed from the chorion over two limited sites corresponding to pre-existing endometrial cushions on the dorsal and ventral wall of each uterine cornu, on either side of the mesometrial attachment. The placenta is therefore bidiscoidal. Structurally the placenta shows progress towards the haemochorial type but, according to the latest work, is of endothelio-chorial type. These details differ considerably from the arrangements in Macroscelididae or any Insectivora; they also contrast markedly with the foetal membranes of lemurs or lorisoids.

Regarding the foetal envelopes and placentation of the primates above the tupaioid level, the conclusions of Hill (1932) are universally adopted. This embryologist distinguished four well defined stages or evolutionary levels, namely:

1. A basal "lemuroid" stage.
2. A transitional or tarsioid stage.
3. An annectant or pithecoid stage.
4. A terminal anthropoid stage.

It is well established that the lorisoidea (both Oriental and African) and all the Madagascan lemurs have a simple diffuse, non-deciduate placenta of an epithelio-chorial variety, resembling that occurring in such ungulates as the pig. There is one exception to the general rule and this occurs in Demidoff's galago (*Galagoides demidovii*) where Gérard (1932) found some unusual and unexpected departures from the usual pattern chiefly, however, in an area of endothelio–chorial placentation surrounded on all sides by the normal diffuse epithelio–chorial condition.

Relation between foetal and maternal elements in the lemurine placenta is produced by the formation of vascularized trophoblastic villi which fit snugly into crypts of the uterine epithelium. Uterine epithelium persists throughout pregnancy and continues to secrete from its glands, at any rate in *Galago* (Strahl, 1899) and *Propithecus* (Turner, 1876) though not in *Daubentonia* (Hill and Burne, 1923). Some minor differences occur in the various genera; thus chorionic villi are simple and unbranched in *Loris*, but in *Nyctice-bus* and *Galago* they are finely branched, while in *Lemur* they take the form of leaf-like folds (for further details consult Amoroso (1961) pp. 185–188).

In the transitional or tarsioid stage, as represented by *Tarsius*, a fundamentally distinct pattern of placental development takes place aligning it with the higher or simian primates rather than with the lemurs, i.e. the placenta becomes localized, disc-like, deciduate and haemochorial in structure. In the early differentiation of the chorion and the manner in which it becomes vascularized, *Tarsius* approaches the higher anthropoids. The allantois, a membrane which in lower mammals and lemurs lies within the chorion and has a vesicular structure, in *Tarsius* grows out from the embryonic gut as a single slender diverticulum into a solid (mesodermal) connecting stalk that develops very early and archors the embryo to the inside of the chorion. On the other hand *Tarsius* presents primitive features in the manner of formation of another foetal membrane, the amnion. Instead of the amniotic cavity being produced as it is in monkeys, apes and man, by cavitation in the inner cell mass, it is formed by the union of folds growing up from the periphery of the embryonic area – as in lemurs and lower mammals.

The pithecoid stage of placental differentiation is found in both New and Old World monkeys. In both, implantation is superficial and so a decidua capsularis or encapsulating layer around the implanted ovum is not developed. The chorion, precociously advanced, is at first shaggy all over, from the rapidly developed and early vascularized branching villi, but is soon restricted to two discoidal areas (with rare exceptions to a single disc) one attached to the dorsal the other to the ventral uterine wall. These discs are haemochorial in structure. The precocious proliferation of chorion from trophoblast is more evident in Old World monkeys than in those of the New where, according to Hill (1932), the placental development is slow and cumbrous so as to reach its full functional capacity, judged on structural considerations, only late in pregnancy. In catarrhines, on the contrary, by acceleration and abbreviation of developmental processes, full function is attained as soon as the foetal circulation is established.

As already noted, the amniotic sac differs from that of *Tarsius* in developing by cavitation of the inner cell mass into an amnio-embryonic vesicle, the floor of which constitutes the ectoderm of the embryonic disc, while the walls and roof are from the amniotic ectoderm. The allantois and body stalk are as in *Tarsius*.

The anthropoid or hominoid stage is very similar to the pithecoid especially the catarrhine representatives, but shows even more pre-cocity in placental elaboration and in the development of a laby-rinthine system of blood-spaces occupied by maternal blood within the trophoblastic layer. A single disc-like placenta, sub-divided into lobate cotyledons is usual. Implantation takes place whilst the blastula is still relatively undifferentiated, burrowing deeply and rapidly into the maternal tissues by the erosive action of the outer trophoblastic layer of the blastocyst (for further details see Wis-locki (1929), Hill (1932) and Amoroso (1961)).

15

Early History of the Primates

Palaeocene and Eocene Lemuroids and Tarsioids

Palaeocene Primates

As there is a dearth of indisputably tupaiid fossils, the original transition from insectivorous ancestral forms to undoubted primates cannot at present be determined; except insofar as it must have occurred at the end of the Cretaceous period. This is certain because, during the middle and late phases of the Palaeocene (basal Eocene) i.e. the dawn of the Tertiary Epoch which occurred around 65 million years ago, undoubted primates were already in existence. Among the wealth of fossils dating from the Palaeocene, it is not always easy to determine whether the owners were surely referable to the primate order or not; some families once considered

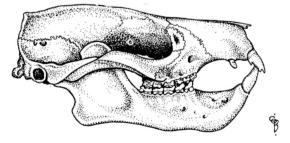

FIG. 70. Restored skull, from the right side, of *Plesiadapis*, a Palaeocene prosimian, based on European and North American fragments. Redrawn from Simons (1964).

Primates are no longer so regarded. There is no question, however, that the behavioural changes and physiological adaptations, as judged by dentition, involved in the adoption of a frugivorous or leaf-eating diet in an arboreal environment, as opposed to a formerly insectivorous menage, took place at this time. Relevant data in connection with this metamorphosis have been reviewed by Szalay (1968) who notes the strong dental similarity between *Purgatorius ceratops* of the late Cretaceous and *P. unio* of the Palaeocene and suggests that the genus is possibly affined to the prosimian family Paromomyidae. This family and the Plesiadapidae are the best known groups of Primates and represent the most primitive of known prosimians, but each has specializations of its own that indicate they represent side lines from the main evolutionary track leading to prosimians of modern type. Paromomyids were represented in the Middle and Late Palaeocene of North America, but the Plesiadapidae were abundant in Europe and North America, surviving in Europe until the early Eocene, with some genera (e.g. *Plesiadapis*) represented on both sides, indicating that, at this date, the two continents were connected by land and enjoyed a tropical or at least a warm temperate climate.

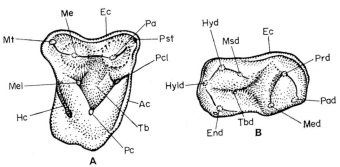

FIG. 71. Second upper molar crown (A) and second lower molar crown (B) of an Upper Palaeocene primate, *Palaechthon*. In both teeth the buccal edge is at the top and the anterior border points to the right.

Ac.	anterior cingulum	Mt.	Metastyle
Ec.	buccal cingulum	Pa.	Paracone
End.	entoconid	Pad.	paraconid
Hc.	hypocone	Pc.	protocone
Hyd.	hypoconid	Pcl.	paraconule
Hyld.	hypoconulid	Prd.	protoconid
Me.	metacone	Pst.	parastyle
Med.	metaconid	Tb.	trigone basin
Mel.	metaconule	Tbd.	talonid basin
Msd.	mesoconid		

Simplified and modified from Szalay

The remarks made above apply with even greater force to the North American family Carpolestidae whose name signifies "fruit-stealers" – an attribute based on the peculiar dentition, which was adapted to opening seeds, berries and other fruit and possibly stripping bark. The hindmost lower premolar had a greatly enlarged and compressed crown with a serrated edge and shows convergence towards the conditions in the multituberculates (Chapter 1, p. 3) as well as in the modern Australian rat-kangaroo *Bettongia*, which subsists on a diet of coarse grass, bark, roots and suchlike fibrous matter. Carpolestids comprise 3 genera that apparently form a succession extending temporally from the Middle Palaeocene to the Early Eocene when they became extinct. These, with the two previously mentioned families form examples of what is known as dead-end evolution, for they left no certainly accredited descendants (see especially McKenna (1966)).

EOCENE PRIMATES

The land connection joining Europe to North America, via Greenland, persisted throughout the Eocene, a matter of some 22 million years. It is not surprising, therefore, that several genera of prosimian primates are represented in both areas. Moreover, by the Middle Eocene it is evident that the tarsioids and lemuroids had become differentiated from one another and from the basal ancestral stock (Simons, 1961a).

As many as 45 probably valid genera of Eocene primates (arranged in 6 families) have been listed by Simons (1963) besides late surviving members of the Palaeocene families referred to above. The latest representative of these was *Phenacolemur jepseni* of the Lysite beds of Wyoming (Middle–Early Eocene). The newcomers are, in the main, distinguished by their greater brain volume compared with body bulk, as witnessed by the earliest known endocast, that of the tarsioid *Tetonius* of the Early Eocene of Wyoming, which is characterized by its large olfactory lobes, small frontal lobes and voluminous temporal and occipital lobes (Radinsky, 1967).

Of the six families of Eocene Primates recognized by Simons (1963) one, the Microsyopidae, has only recently been recognized as of primate status (McKenna, 1960). Its constituent genera, *Cynodontomys* (Early Eocene of North America), *Microsyops* Middle Eocene of North America), *Alsaticopithecus* (Middle to Late Eocene of Europe) and *Craeseops* (Late Eocene of North America) had been previously confused and combined in the Insectivore

Fig. 72. *Platychaerops richardsoni*. A Lower Eocene plesiadapid lemur from Herne Bay, Kent, England. Palatal view of the maxilla with right P⁴, M², M³, and left M² and M³.

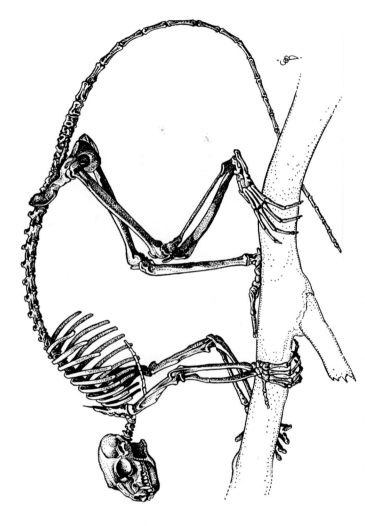

FIG. 73. Restored skeleton of *Smilodectes gracilis*, an Eocene, lemur-like prosimian, from Wyoming. Redrawn from Simons (1964).

family Mixodectidae – so intertwined are these fossils. The New World genera appear to represent a temporal succession characterized by increasing molarization of premolars. In size the microsyopids ranged from that of a small cat to a large dog and their affinities seem rather with the Palaeocene plesiadapids and Eocene Adapidae than with other families.

The large and well known Eocene family Adapidae consists of 9 genera of definitely lemuroid primates, three (*Notharctus, Pelycodus* and *Smilodectes*) from North America, the remainder being European. The oldest were *Notharctus* (Early to Middle Eocene) and *Pelycodus* (Early Eocene). The European forms mostly, like *Adapis*, ranged from the Middle into the Late Eocene. Several of these genera are known from almost complete skeletons, so their structure is quite well authenticated. Notable are the studies of Gregory (1920) on *Notharctus*, Stehlin (1916) on *Adapis* and Gazin (1958) on *Smilodectes*.

The degree to which these adapids had surpassed structurally the level attained by Palaeocene forms like *Plesiadapis* is indicative of the relatively rapid rate of early primate evolution. The New World forms in particular resembled living Malagasy lemurs in their size and general structure. Compared with the small brained, long snouted *Plesiadapis*, the skull of *Smilodectes* exhibits an enlargement of the frontal region of the brain case and a forward shift of the optic axes; combined with the elongated hind limbs an impression is gained of a lemur somewhat resembling the modern *Propithecus*. Neither *Smilodectes* nor *Notharctus*, however, had claims to be ancestral to surviving lemurs, a role that was more probably played by the European forms such as *Adapis* or *Protoadapis*, unless their ancestors were already inhabitants of Africa (Simons, 1964).

The European adapids showed much variability and it may be possible that some, such as *Pronycticebus* and *Anchomomys* were the precursors of the lorisoids, but dental similarities between these and undoubted lemuroids like *Pelycodus* and *Protoadapis* render taxonomic separation difficult and unjustified.

Pronycticebus was originally so named by Grandidier (1904) because of certain features which suggested lorisoid affinities. Gregory (1920) and Abel (1931) removed it to the Tarsioidea, but a restudy by Le Gros Clark (1934) showed that, in the anatomy of its auditory region, it was more lemuroid than tarsioid. Simons (1962) however had discussed possibilities of a relationship with such Miocene lorisoids as the East African *Progalago*, so we are now back to Grandidier's stand.

Among the remaining Eocene primates not so far considered are 30 genera that may be regarded as tarsioids. In the past they have been relegated to a single family, Anaptomorphidae (or more correctly Microchoeridae) (e.g. Hill (1955)), but divisible into 5 subfamilies as suggested by Simpson (1940). Some of the genera show such close structural resemblance to the living *Tarsius* that separation from the Family Tarsiidae is scarcely warranted, so that Simons (1963), for example, includes them under this family as the subfamily Necrolemurinae. At the same time Simons (1963) promotes the former sub-family Omomyinae to full family rank, so that the Eocene tarsioids are now grouped as follows:

1. Family Anaptomorphidae, exclusively North American, of Early and Middle Eocene age; examples: *Anaptomorphus, Tetonius.*
2. Family Omomyidae, occurring in North American, European and Asiatic deposits throughout the Eocene; examples : *Omomys* – Early to Middle Eocene, N. America; *Periconodon* – Middle Eocene, Europe; *Hoanghonius* – ? Late Eocene, China.

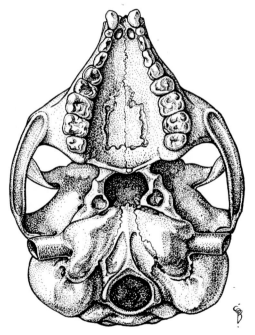

Fig. 74. Basal aspect of the cranium of *Necrolemur*, a European Eocene tarsioid. $\times \frac{2 \cdot 5}{1}$ Redrawn from Simons and Russell.

3. Family Tarsiidae, subfamily Necrolemurinae; exclusive to Middle and Late Eocene of Europe; examples: *Necrolemur* – Middle to Late Eocene; *Pseudoloris* – Late Eocene; *Microchoerus* – Late Eocene.

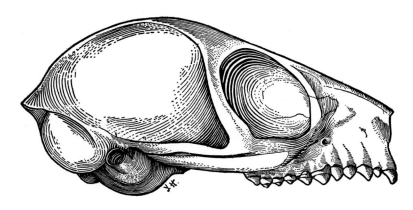

FIG. 75. Lateral view of the cranium of *Necrolemur antiquus*. $\times \dfrac{2\cdot 5}{1}$

It is admittedly difficult to decide, in dealing with Eocene fossils, on assignment of a particular specimen to a lemuroid or a tarsioid assemblage; in fact, Simpson (1955) categorically stated that meaningful distinction between them cannot be found. Much the same sort of bandying about has been the fate of *Necrolemur* and its allies as we have encountered in discussing *Pronycticebus*. Thus Abel (1931), Gregory (1922), Simpson (1940) and Clark (1959) all place them with Tarsioidea; Simpson in fact stating that, on the basis of dentition, *Pseudoloris* is so much more like *Tarsius* than any other Palaeocene or Eocene primate that it should be placed in the Tarsiidae. On the contrary Hürzeler (1948), in a comprehensive study of the Necrolemurinae, concluded that they should constitute a subfamily of Lemuroidea as he saw no meaningful resemblance to *Tarsius*. More recent studies by Simons (1961a) have resulted in their return to the Tarsioidea as a subfamily of Tarsiidae. Most of the genera of Eocene so-called tarsioids are insufficiently known anatomically, especially in their cerebral and middle ear anatomy, to be certain of their affinities, but this statement cannot apply to *Necrolemur* or to some, at least, of the Omomyidae (e.g. *Hemiacodon*, whose facial bones and much of the hind limb skeleton are known.)

None of the Necrolemurinae, with the possible exception of

Pseudoloris, could have been ancestral to the living tarsiers, because they already possessed too specialized a dentition. Their presence in the European Eocene, however, serves as a document for the beginning of one of the major types of surviving Primates.

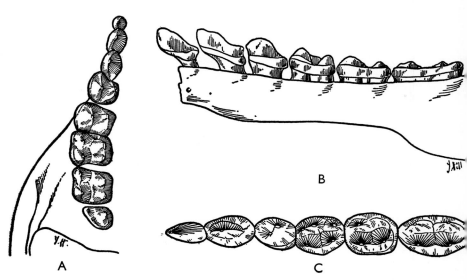

Fig. 76. Dentition of *Necrolemur.* A. Upper dental series in occlusal view; B. mandibular series, premolars and molars in lateral view; C. the same, in occlusal view. From Hill (1955).

THE OLIGOCENE

A few lowly Primates allied to Eocene genera lingered on into the Oligocene. Two of these, *Macrotarsius* (Clark, 1941) from Montana and *Rooneyia* (Wilson, 1966) from Western Texas appear to be omomyid tarsioids. *Rooneyia* is particularly important as it is represented by a well preserved cranium containing a natural endocast. The endocast indicated that cerebral hemispheres were at a level of development comparable with *Tarsius* and that the area concerned with visual reception was highly developed and the olfactory region reduced (Hofer and Wilson, 1967).

On the evidence of some limb bone fragments from the Early Oligocene beds of the Isle of Wight, Day and Walker (1969) have attributed a humeral fragment and a pedal middle phalanx to the lemuroid *Adapis magnus.*

Although fossil representatives of other (more advanced) primate groups had made their appearance in the Oligocene, the above are the last heard of the prosimians until the advent of some lorisoid genera in the Miocene of East Africa and *Indraloris* of the Pliocene of India.

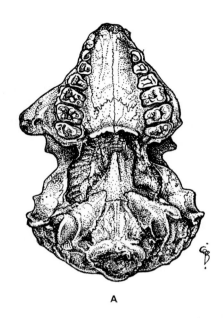

A

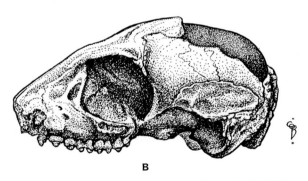

B

FIG. 77. Skull and endocranial cast of the fossil Oligocene tarsioid *Rooneyia viejaensis*. Drawn from a plaster cast kindly supplied by Prof. F. A. Wilson. A. from the palatal aspect; B. from the left side. $\times \frac{1}{1}$

16

Madagascar: The Effects of Isolation

The true lemuroids, descendants presumably of the Adapidae, are today confined to Madagascar and some small neighbouring islands (e.g. the Comoros). Questions arise, therefore, concerning the date and route of their arrival. There are, at present, no answers to these questions, insofar as mainland Africa has thus far yielded no fossil adapids. The only surviving African prosimians and the only fossils so far recovered are all of lorisoids – either galagos or pottos – which did not succeed in migrating to Madagascar (see next Chapter).

Madagascar became separated from Africa by the formation of the Mozambique channel in late Primary or early Secondary times, but subsequent elevation of narrow bridges permitted access to the island of several groups of primitive mammals until well on in Tertiary times. Final separation must have taken place in late Eocene or at most earliest Oligocene times, for immigration of adapid ancestors of the lemurs was achieved before any simian competitors had evolved; these were evolving in Africa during the Oligocene. The Comoro Islands represent a relic of the latest land bridge (Millot, 1952). The adapids had manifestly gained entry with other early mammals from Europe into Africa during the Upper Eocene by a land bridge to the west of the Tethys Sea, either at the site of the present Straits of Gibraltar or via the Sicily–Malta–Tunis bridge (Moreau, 1952).

The fact that lemurs are the only primates to develop in Madagascar supports the view that their ancestors arrived prior to the evolution of more advanced monkey-like primates that developed in forested areas of Africa. Persistence in the island of other primi-

FIG. 78. Ring-tailed Lemur (*Lemur catta*). From Hill (1953).

tive mammalian types (especially insectivores, rodents and the fossa (*Cryptoprocta*)) likewise lends support to the view that these, along with diurnal lemurs, escaped competition with advanced predators and competitors that soon afterwards appeared in Africa. As Buettner-Janusch (1963) remarks "the discovery of fossil tenrecs in Kenya may presage the discovery of fossil lemurs in the right time zones to document this fully".

Having once gained sanctuary in Madagascar, the ancestral lemurs underwent rapid evolutionary deployment unhindered by simian competition or predator hazard. They rapidly exploited every potential ecological situation, and niches were extremely varied from rain-forest, montane forest, savannah, rocky outcrops, marshes, swamps (with their reed beds) and even, according to some workers, an aquatic environment (Sera, 1938, 1947, 1950). The result was a wealth of speciation and subspeciation unheard of elsewhere (see also discussion in Hill (1953) pp. 311–316).

Some forms, by keeping to nocturnal habits in a forest habitat, remained remarkably primitive (e.g. the Mouse Lemur, *Microcebus*

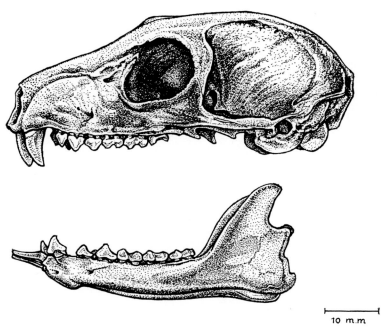

10 m.m

Fig. 79. Skull, with mandible, of a true Malagasy lemur (*Lemur* sp.) Note tympanic bulla, large upper canines and procumbent anterior mandibular teeth. $\times \frac{1}{1}$

and Dwarf Lemurs, *Cheirogaleus* and *Phaner*). One branch, also retaining a nocturnal forest habitat, became adapted, like some of the early plesiadapids and carpolestids, to a rodent-like existence, chiselling bark and wood to obtain wood-boring beetles; these were the family Daubentoniidae (aye-ayes) so specialized that, at one time they were thought to be descended directly from plesiadapid ancestors. But so frequent was this type of adaptation, occurring as it did, in several independently evolving primate lines, that the dental and other similarities between them are nowadays interpreted as the result of parallel evolution.

Significant, however, for further evolutionary radiation, was the adoption of diurnal activity. This led to the emergence of such highly successful forest forms as the members of the genus *Lemur* and the two genera *Propithecus* and *Indri* of larger creatures – the last being the largest extant member of the suborder. On the other hand *Lepilemur* and *Avahi* remained nocturnal and are the smallest of their respective groups.

Hapalemur has become specially adapted to a life in the reedbeds that border the larger lakes of the island.

In the early Pleistocene these lakes were more extensive than they are today. From deposits around their present shores has been recovered a wealth of skeletal material of subfossil lemuroids, some of them heavily built and of gigantic proportions compared with surviving arboreal forms. Such large creatures must have forsaken their forest home to have adopted a terrestrial mode of life or even, as Sera (1935, 1938, 1947, 1950) suggests possibly in the case of *Palaeopropithecus*, a swamp dwelling or semiaquatic habit.

At least 14 species of extinct lemurs are known from Madagascar. One was an aye-aye much larger than the surviving species. Two were large members of the genus *Lemur*. There were 8 species, comprising 6 genera (*Neopropithecus, Mesopropithecus, Archaeoindris, Palaeopropithecus, Archaeolemur* and *Hadropithecus*) more or less closely affined to the existing Indriidae, but mostly larger than the present day giant of the family *Indri*. Several members of this group, e.g. the first two mentioned, had shortened faces and were even more monkey-like in appearance (and probably also in respect of ecological adaptation) than any of the surviving genera, indicating that latent trends to evolve what was virtually a monkey were manifested in more than one primate lineage.

Finally the real giants were the 3 species of the genus *Megaladapis, M. madagascariensis, M. grandidieri* and *M. edwardsi*, listed in order of increasing size, the last mentioned attaining the size

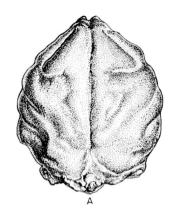

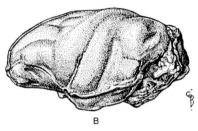

Fig. 80. Endocranial cast of the Crowned Sifaka (*Propithecus verreauxi coronatus*), an indriid lemur, from the upper (A) and lateral (B) aspects to show the large size of the cerebral hemispheres, almost completely masking the cerebellum from above, also the richness of the cerebral convolutions. $\times \frac{1}{1}$

of an Alsatian dog. These must have been clumsy, heavily built animals having massive limb bones. They were evidently terrestrial in habit, foraging along the borders of swamps and vulnerable to their enemies, though one view is that they were slow moving arboreal climbers analogous to the marsupial koala (*Phascolarctus*).

The question arises as to what these enemies were in the absence of large carnivores. Although there may have been minor climatic and ecological changes affecting the island, the balance of blame must be laid upon direct or indirect human activity in prehistoric times (between 1000 and 2000 years B.C.) The evidence has been ably discussed by Walker (1967) who also offers conclusions regarding locomotor habits of the extinct forms.

17

Emergence of the Lorisoids

Surviving lorisoids are confined to tropical and subtropical forested zones of the African continent and southern Asia. They did not reach Madagascar. Unquestionably they comprise "a valid phylogenetic taxonomic unit of rather closely related groups" (Simpson, 1967). Many include them in a single family Lorisidae, divided into two well defined subfamilies, Lorisinae (lorises and pottos) and Galaginae (galagos, bush babies); but some have promoted each of these to family rank.

Although retaining, in common with the Lemuroidea, many primitive features and remaining phylogenetically at the prosimian level, the lorisoids have developed a number of specializations; especially

(i) the great expansion of the orbits to conform to enlarged eyeballs and correlated facial modifications, without much restriction of the nasal fossae.

(ii) the formation of a complete bulla tympanica incorporating laterally the ectotympanic and an inflation of the mastoid area.

(iii) reduction of incisors to $\frac{2}{2}$ and premolars to $\frac{3}{3}$. The lower canine is incisiform, sharing with the incisors the function of a dental comb; $P_{\overline{2}}$ is caniniform. These dental specializations are shared with the Lemuroidea. Molars are essentially primitive, but $M^{\underline{1-2}}$ bear large hypocones and $M_{\overline{1-3}}$ lack distinct paraconids.

(iv) pollex and hallux are widely divergent, II is reduced and IV the longest digit, opposing I in grasping.

Fig. 81. Slender Loris (*Loris tardigradus*). Photo W. C. Osman Hill.

The living lorisoids, though all are nocturnal, fall naturally into two groups based on locomotor habits. In one, the Galaginae, locomotion is by rapid saltation. In adaptation to this the hind limbs are elongated and muscularly powerful; the tail is long and ears are large, membranous, highly mobile and capable of folding concertina wise. In contrast to this the lorises and pottos are slow moving hand over hand climbers; their hands and feet are capable of very powerful grasp; their tails are reduced or absent and ears small and rounded. Some cranial and dental features also distinguish the two groups; these are useful in analysing fossil fragments. They are tabulated by Simpson (1967) but with cautious remarks as to their being sufficiently clear-cut to serve in diagnosis – except for the greater tendency to molarization of P $\frac{4}{4}$ in galagos.

Galagos are restricted to Africa where, however, their range is very wide; but there is no evidence that they were ever represented outside Africa.

Four species of living lorisines are commonly admitted, each in a separate genus: *Loris* (confined to South India and Ceylon), *Nycticebus* (North eastern India, South eastern Asia and Malaya), *Arctocebus* (West Africa) and *Perodicticus* (West and Central Africa

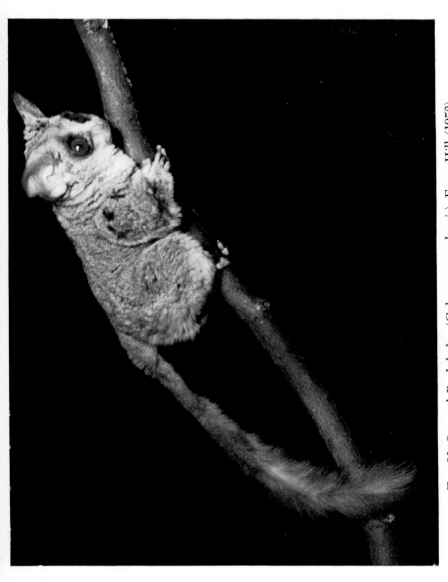

FIG. 82. Senegal Bush-baby (*Galago senegalensis*). From Hill (1953).

as far as the Rift Valley or beyond). *Loris* and *Arctocebus* are much alike, so are *Nycticebus* and *Perodicticus,* but these similarities may be due to convergence. Simpson (1967) suggests the possibility that two ancestral lineages first became differentiated and that in each line a geographically isolated, vicarious pair of species and genera later evolved in Africa and Asia.

Fossil lorisoids of both locomotor types have been recovered from Miocene deposits in East Africa, but from Asia an $M_{\overline{2}}$ from Pliocene and a mandibular fragment from Miocene deposits of the Simla Hills described by Lewis (1933) and Tattersall (1968) as *Indraloris lulli* are the only evidence of their existence in the past. Lewis believed the tooth could have been derived from that of an adapid ancestor and to have served as precursor of *Nycticebus.*

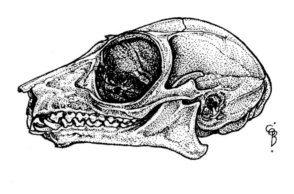

├────────────┤
10 m.m.

Fɪɢ. 83. Skull, with mandible, of a Galago (*G. alleni*) showing proportions of neuro-cranium and facial structure, also the large size of the orbit.

Of the East African fossil lorisoids two species of *Progalago, P. dorae* and *P. songhorensis,* are found in the Miocene of Kenya; a new genus *Komba* has been proposed by Simpson (1967) for two other galagine species originally referred by Le Gros Clark and Thomas (1952) to *Progalago.* These are *K. robustus* and *K. minor,* both likewise from the Miocene of Kenya. Leakey (1962) has reported from the Miocene of Napak, Uganda, a fossil galagine allied to the West African Needle-clawed Galago (*Euoticus elegantulus*); this he has called *Mioeuoticus bishopi.* Finally a supposed fossil potto from the

Miocene of Kenya has been identified by Simpson (1967) as *Propotto leakeyi*.

All these African Miocene lorisoids have been reviewed by Simpson (1967) who concluded that none has any special resemblance to any one of the Recent genera or species, so that he considers names like *Progalago* and *Mioeuoticus* misleading. Some bear features not matching in any of the surviving forms; but at least the main dichotomy between a galagine and a lorisine lineage had become established in the Miocene though all, except *Propotto*, betray a mixture of resemblances to Recent members of both subfamilies, but there are no characters in which any of them can be considered more primitive than the Recent genera or in which the latter are more specialized than the Miocene forms. The East African fossils, therefore, are of little help in closing the evolutionary gap between the two most likely European adapid forerunners, the Eocene *Pronycticebus* and the Oligocene *Anchomomys*.

More recent researches by Walker (1969, 1970) have shown that *Propotto* is a fruit bat while *Mioeuoticus* is not separable from *Progalago*. All the known East African fossil lorisoids, therefore, are galagine; a view which is supported by the acquisition of postcranial skeletal material.

18

Tarsius—A Living Fossil

An even greater gap in the geological record separates the Eocene and Oligocene tarsioids from the surviving *Tarsius*, than we have encountered in the lorisoid record. *Tarsius*, therefore, is aptly described as a "living fossil" and consequently is regarded as a scientific curiosity of immense importance.

Tarsiers are rat-sized Primates, so called by Buffon (1765) on account of the elongated tarsal region of the foot. They are today confined to certain islands of the Indonesian and Philippine Archipelagos, namely Sumatra, Bangka, Belitung (= Billiton), Karimata and the South Natuna Islands, Borneo, Celebes, Great Sanghi Island and the following islands in the southern Philippines : Samar, Leyte, Bohol and Mindanao. There is no evidence of occurrence on the Asiatic mainland either in Recent or past geological times, nor are they known from Java. Three species may be recognized, a western *T. bancanus;* a Philippine *T. syrichta* and a Celebesian *T. spectrum.* These differ in some minor pelage characters, some cranial features and in some important structural peculiarities of the caudal integument (Hill, 1953; Sprankel, 1965).

On the whole tarsiers give the impression of prosimian status recalling galagos in their soft, woolly pelage, grotesquely large eyes, expanded membranous ears and elongated hind limbs, adapted for saltatory locomotion in an arboreal environment. They are, however, at once set apart by the absence of a naked rhinarium, for their shortened muzzles are characterized by the presence of completely rounded nostrils widely separated by an internarial septum covered by normal hairy skin, which also surrounds the somewhat outwardly directed nostrils up to their margins (haplorhine condi-

FIG. 84. Philippine Tarsier (*Tarsius syrichta carbonarius*). Living specimen photographed in the typical resting position with hands and feet bunched, closely clasping a vertical branch, and the tail used as a support. Note the goggle-like eyes, large bat-like ears and the small rounded, widely separated nostrils (photo E. Walker).

tion, in contrast to the prosimian strepsirrhine condition). This narial condition is brought about developmentally by the complete fusion of the mesial and lateral nasal processes of the embryonic visage, the mesial process forming, in addition to the broad inter-narial septum, a median part of the upper lip or philtrum in place of the median cleft lip of the lorisoids and lemurs. Coincident with this the upper lip is freed, being no longer tethered to the gum, rendering it more mobile and hence available for expression of emotions. Large glandular accumulations affect the lateral parts of the upper lip (Montagna and Machida, 1966) and also the lower lip (Sprankel, 1970).

The tail of the tarsier is unique among Primates; its closest re-

semblance is to that of *Ptilocercus,* but the resemblance is superficial only. It is very long and slender and the major part of its
length is virtually naked; the base, however, is hairy and there is a
terminal sparse tuft varying with the species. The organ is adapted
to the peculiar resting pose of the animal on a twig and to affording
a purchase in leaping from the resting position to a new situation.
In fact, like the tail of the kangaroo or a jerboa, it serves as the
third limb of a tripod. A specialized friction surface occurs in the
proximal part of the tail ventrally in *T. syrichta* and *T. bancanus,*
but in *T. spectrum* this is modified further to present a rasp-like
system of tooth edged scales (Hill, 1953; Sprankel, 1965).

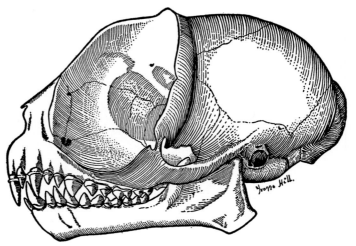

FIG. 85. *Tarsius.* Lateral view of the cranium and mandible. From Hill
(1955).

Limbs too show adaptations to the mode of life. Forelimbs are
short and relatively weak but the hind limbs are long and muscular,
being held, when at rest, flexed on the abdomen after the fashion
of a frog. The ankle region, as in the frog also, is elongated, the
elongation affecting especially the calcaneus and the navicular, the
talus being normal in shape, except for having a longer neck.
Another feature, unique among Primates, is the fusion of the lower
end of the fibula with the tibia; doubtless this gives greater stability
to the ankle joint. Sudden extension of the hind limbs gives tremen-
dous impetus to the body in leaping from one position to another.
The mechanics of the limb, with special reference to the hip joint
and its musculature have been studied in detail by Grand and
Lorenz (1968).

Further adaptations affect the hands and feet, both of which are virtually naked; terminal pads of fingers and toes are expanded into round, flattened discs which function both as tactile pads and, by enlarging the area of contact, serve to increase stability, preventing slipping of the grip. Nails are small and scale-like, except on the index and middle toes of the foot, both of which (instead of only the index as in lemurs and lorisoids) are provided with upstanding conical claws for use in combing the fur. In the digital formula of the hand, the middle finger is longest, not the fourth (cf. lemuroids).

In the skull the most striking feature is the enormous size of the orbits, which are partly closed in behind. They face directly forwards compressing the nasal fossae. The face is short and wide and the brain case broad and globular, with the foramen magnum placed well forwards and downwardly faced. There is a prominent tympanic bulla and a short tubular external auditory passage. The slender mandible has its rami meeting in V-fashion; ascending rami are broad and short.

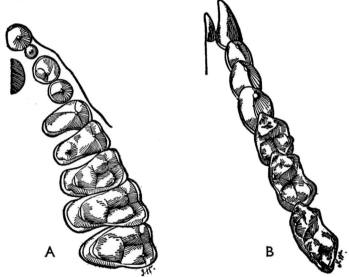

Fig. 86. *Tarsius*. Dentition. A. left maxillary series in occlusal view; B. right mandibular series in occlusal view. $\times \dfrac{4}{1}$

The dental formula is $I\frac{2}{1}$; $C\frac{1}{1}$; $P\frac{3}{3}$; $M\frac{3}{3} = 34$. Upper central incisors are large and closely approximated, the laterals small. Lower incisors are not procumbent as in Prosimii, but small and erect. Canines, though projecting are rather small. $P\underline{2}$ and $P\underline{3}$ exhibit a well developed cingulum on the lingual side, but lower premolars are small, with simple conical crowns. Upper molars are tritubercular and the lower ones tuberculo–sectorial.

In the alimentary tract the stomach is simple and globular, the small intestine arranged in simple loops with the large intestine forming a "U" around them. Caecum and colon are about equal in length and show no flexures or coils or sacculations. The peritoneum has a simple arrangement.

The large eyeballs have a protuberant cornea, iris extensive with the pupil reducible to a horizontal slit in bright lights; the retina is a rod-retina but a macula and fovea are present (see Chapter 3). Eyeballs are capable of little movement, but the loss is compensated for by the extreme rotatory power of the head (through 180°).

In the genital tract the uterus is bicornuate, as in the lemurs and lorises, but placentation differs markedly in being discoidal, deciduate and of haemochorial structure as in monkeys.

In sum, therefore, *Tarsius* is an enigma comprising, as it does, a complex mosaic of anatomical features, some primitive or generalized and some specialized. The specializations are largely adaptations to its special mode of life and appear to be of ancient origin inasmuch as the limb structure had already been acquired by the Eocene tarsioid *Hemiacodon*. In other respects *Tarsius* is the most primitive existing Primate – apart from the tupaioids. It retains many primitive traits in common with the prosimians, lacks most of their specialization through having arrived at others of its own not seen in the prosimians or, if shared with them, they have been evolved by parallelism. Notwithstanding, *Tarsius* is also annectant with the simians for example, having lost the primitive rhinarium and developed a monkey-like visage, having developed an incomplete partition between orbit and temporal fossa and, above all, having improved on the relationship between the foetal membranes and maternal tissues in its ontogenetic development.

19

The History of Primates in the New World

South America became isolated from North America in the Eocene or even earlier. Before this separation occurred Primates must have gained access from the northern land mass to the southern. The pioneers must have been of prosimian status, but no fossil evidence of these forebears of the platyrrhine monkeys has so far come to light but, like marsupials in Australia and lemuroids in Madagascar, the newcomers found all ecological niches free and consequently rapidly adapted themselves to fit into them.

We have no definite clue as to which of the North American Eocene prosimians extended its range into South America, but we can be sure that among these was the ancestral group or groups. The most southerly prosimian so far recovered from the New World is the tarsioid *Rooneyia* from the early Oligocene of Texas, but by this time one or other of its precursors must have already gained the southern land mass.

The most northerly ranging of surviving New World monkeys is the Mexican race of the Black-handed Spider Monkey (*Ateles geoffroyi vellerosus*) but this belongs to perhaps the most highly evolved genus of the whole family, so that it offers no assistance in plotting the deployment of Primates in the southern contir~~~t being~ on the contrary, probably a late emergent, colonizing Cen the south, subsequent to the reunion of the two c

The most ancient fossil Primates so far reco America are of late Oligocene to Miocene age an pertaining to the genera *Homunculus*, *Stirtoi Cebupithecia* and *Neosaimiri*, and all are comp types. Together they present a spectrum of forms closely paralleling, if not in fact ancestral to, their living counterparts. They have been critically reviewed by Stirton and Savage (1951) and more recently by Hershkovitz (1970) who considers that, during the Tertiary, platyrrhine monkeys were more diversified than they are today in terms of generic and subfamilial categories, besides being more widely distributed geographically. Thus there appear to have been centres of evolutionary radiation located in Patagonia, north-western South America and the Greater Antilles.

Homunculus of Patagonia proves to be an advanced platyrrhine not specially related to any living genus, though suggestions of affinity with *Alouatta* have been made. *Dolichocebus* from the Upper Oligocene of Argentina is said by Hershkovitz to be closer to *Homunculus* than any other genus, but its small size, cranial form and dental formula suggests an affinity with the marmosets. From the La Venta formation of Miocene age of Colombia are derived the genera *Stirtonia, Cebupithecia* and *Neosaimiri.* The first of these was originally referred to *Homunculus,* but Hershkovitz has elevated it to separate generic and subfamily status. *Cebupithecia,* known from a complete skull, betrays features especially in the dentition aligning it with the modern Sakis (*Pithecia*), though this is denied by Hershkovitz, who attributes the condition to post-humous distortion and regards this also as representing a distinct independently evolved subfamily. On the other hand *Neosaimiri* closely resembles its modern counterpart *Saimiri,* which is a relatively primitive platyrrhine.

As Simons (1963) remarks, published figures of the crania of *Homunculus* and *Dolichocebus* suggest clearly that the bun-shaped cranial form had already been attained in Miocene times, but the frontal lobe expansion was not as advanced as in modern Cebidae of comparable size.

From sub-Recent cave deposits in Jamaica an imperfect sub-fossil primate mandible has been given the name of *Xenothrix mcgregori* (Williams and Koopman, 1952). No other fossil monkey has been recovered from any Caribbean island and no living forms are found there. Its dental formula, $I_{\frac{2}{2}}$;$C_{\frac{1}{1}}$; $P_{\frac{3}{3}}$; $M_{\frac{2}{2}}$, is the same as that of the marmosets, but in size it approaches a small saki. Hershkovitz considers its geographical isolation and distinctness from known forms entitles it to a separate family rank (Xenothricidae).

Recent platyrrhine monkeys show a remarkably wide variation in structure, but all are in agreement in the undermentioned features which distinguish them from Old World monkeys:

Fig. 87. White-shouldered Marmoset (*Callithrix humeralifer*). Male carrying infant dorsally. Photo L. Bowling; Yerkes Regional Primate Center.

1. The nostrils are widely separated and open laterally rather than downward (platyrrhine).
2. The pollex is not opposable, but flexes in line with the other manual digits.
3. There are no cheek pouches.
4. The external auditory meatus is not prolonged as a bony tube, the tympanic bone remaining ring like. A tympanic bulla is present.
5. There are no ischial callosities.
6. In the brain box the alisphenoid meets the parietal, the latter also articulating with the orbital plate of the malar bone.
7. Orbito–temporal foramen large.
8. Premolars three in number above and below.
9. No sigmoid flexure on descending colon.
10. Caecum not narrowing appreciably towards its apex and usually bent in form of a hook.

The New World Primates together constitute the Infraorder Platyrrhini and Superfamily Ceboidea. They are taxonomically arranged in two very distinct families, the Callithricidae (or Hapalidae) marmosets and the Cebidae, which appear to be the result of two separate evolutionary lines. Their disparities are, however, to some extent bridged by the peculiar genus *Callimico* which resembles a large marmoset, but has the dental formula of a cebid.

Family I—Callithricidae—Marmosets and Tamarins

This includes small, long tailed Primates ranging in size from the pygmy *Cebuella* to the squirrel-sized Lion Marmoset (*Leontideus*). The family is characterized by

(i) a dental formula $I\frac{2}{2}$; $C\frac{1}{1}$; $P\frac{3}{3}$; $M\frac{2}{2} = 32$;
(ii) all digits except the hallux, are provided with long, curved claws; the hallux bears a broad, flat nail;
(iii) the cerebral hemispheres are smooth with few convolutions and
(iv) twins normally result at conception; after birth they are carried dorsally by the male parent, who passes them to the female at feeding times only.

There are two main groups, those with normal canine–incisor relationships in the lower dentition (tamarins) and those where lower incisor crowns are elongated to bring them level with the canines (marmosets). Each group contains several genera thus:

		No. of species
Tamarins	Genus *Saguinus*—Hairy-faced Tamarins	16
	Genus *Marikina*—Bald Tamarins	4
	Genus *Oedipomidas*—Crested Bare-faced Tamarins	2
	Genus *Leontideus*—Lion Marmosets	3
Marmosets	Genus *Callithrix*—Hairy-eared Marmosets	
	(Subgenus *Mico:* Naked-eared Marmosets)	8
	Genus *Cebuella*—Pygmy Marmoset	1

All are adapted to arboreal diurnal life in forest or scrub. Their climbing behaviour is reminiscent of squirrels.

Geographically the family ranges, in suitable terrain, from Panama to south-eastern Brazil. Recent reviews of the family or parts of it are found in Hill (1957) and Hershkovitz (1966a, b, 1968). For the Lion Marmosets see papers by Coimbra-Filho (1969, 1970a, b).

The genus *Callimico,* with a single species, (*C. goeldii*) is essen-

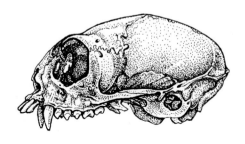

12 mm

FIG. 88. Left lateral view of the cranium and mandible of a Common Marmoset (*Callithrix jacchus*).

tially a marmoset having, like them, clawed digits, but it retains M$\frac{3}{3}$, for which reasons some have classified it in the Cebidae, others in the Callithricidae; Hill (1959) believed it to be a primitive marmoset – an offshoot of the line leading on the Callithricidae before the loss of M$\frac{3}{3}$, but Hershkovitz (1970) considers it the end product of an independent evolutionary radiation and makes of it a separate family, Callimiconidae.

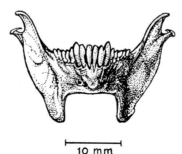

|———— 10 mm ————|

FIG. 89. Frontal view of the mandible of a Common Marmoset (*Callithrix jacchus*), to show the elongation of the incisors.

Family II—Cebidae—South American Monkeys

In this large family the dental formula is I$\frac{2}{2}$; C$\frac{1}{1}$; P$\frac{3}{3}$; M$\frac{3}{3}$ = 36. All digits are provided with flat nails, but the pollex still remains unopposable, though capable of divergence. The tail is long (except in *Cacajao*) and in some genera prehensile. Ears are more or less naked and not usually adorned, as they are in many marmosets, with tufts or pencils of hair. A single young is born after each gestation and carried by the female on her back.

These monkeys occur throughout the tropical forested areas of America from southern Mexico to south-eastern Brazil and northern Argentina, being specially well represented in the Guianan region and the Amazonian forests. Eleven genera are recognized, grouped into 5 or 6 subfamilies thus:

Aotinae	*Aotes*—Douroucoulis or Night monkeys	1 species
	Callicebus—Titis	3 species
Pitheciinae	*Pithecia*—Sakis or Sakiwinkis	2 species
	Chiropotes—Bearded Sakis	2 species
	Cacajao—Ouacaris (pronounced Wakarees)	3 species
Saimirinae	*Saimiri*—Squirrel monkeys	2 species
Cebinae	*Cebus*—Capuchins	4 species
Alouattinae	*Alouatta*—Howlers	5 species

Atelinae	{	*Lagothrix*—Woolly Monkeys	2 species
		Brachyteles—Wooly Spider Monkeys	1 species
		Ateles—Spider Monkeys	4 species

AOTES

Douroucoulis are the only simians adapted to a fully nocturnal habit. Like some of the nocturnal prosimians, they spend the day in hollow trees, sleeping in a hunched quadrupedal posture. The eyes are large and protuberant, adapted for night vision (pure rod retina) and the orbits accordingly enlarged and forwardly directed. The braincase is low, short and broad; a large tympanic bulla is present. The tail is long, well furred with a glandular area ventrally at its base and non-prehensile.

CALLICEBUS

Titis are medium sized diurnal, rain forest forms with soft lustrous pelage, with a non-prehensile tail much longer than the body. Limbs are short, the hind longer than the fore. In the skull the facial part is deep and the angular region of the mandible expanded, recalling the howlers. The voice has a howler-like quality (Moynihan, 1966). For social behaviour see Mason (1966). For review of species see Hershkovitz (1963).

PITHECIINAE (FIG. 90)

The sakis and ouakaris are all larger than the preceding animals and characterized by their long harsh coats and extremely broad noses. The tail is non-prehensile, long in sakis and short in *Cacajao*. Lower front teeth are procumbent and the upper incisors projecting. In *Chiropotes* a large beard is present and the crown hair bouffant, whereas in *Cacajao* the scalp is bald and the bare areas brilliant crimson. The diet in all 3 genera is wholly frugivorous, with appropriate adaptations in the gut.

SAIMIRI

Squirrel monkeys are the smallest Cebid monkeys with characteristic facial appearance with white circumocular discs and a sharply defined black muzzle. The ears are adorned with hairy tufts and the coat short, dense and close fitting. The tail is very long,

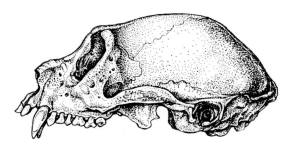

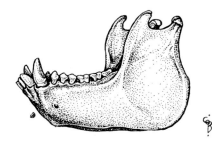

18 mm

FIG. 90. Skull and mandible of a New World monkey—a White-headed Saki (*Pithecia*). Note particularly the cranial sutures and the absence of a bony auditory canal.

closely haired and non-prehensile. Anatomically primitive, the genus betrays a curious mixture of generalized and specialized features, e.g. a simple gut arrangement combined with a complicated peritoneum. A single young is produced at birth, large in proportion to the adult; there is some sexual dimorphism in size. Squirrel monkeys live in large social groups on the forest fringes.

CEBUS

Capuchins are larger and more robust than the preceding and prefer the interior of tropical rain forests. They possess long, closely haired prehensile tails carried rolled or coiled ventrally, whence they derive the trade name of "ring tails". The heads are either rounded or in the case of one species (*C. apella*) somewhat angulated from the presence of tufts on the crown. The brain is large and well convoluted. Capuchins display a high level of intelligence – in advance of many Old World monkeys.

ALOUATTA

Howler monkeys are the largest platyrrhines. They owe their name to their vocal capacities which, in turn, depend on the greatly inflated hyoid bone, which contains an air sac derived from the larynx. The mandible is expanded to protect this arrangement and there are other adjustments elsewhere in the skull. The long, highly mobile tail is prehensile and provided ventrally in its terminal third with a naked tactile pad marked by papillary ridges like a finger tip. The diet is vegetarian, the stomach being capacious, as is the large intestine, although this latter is short.

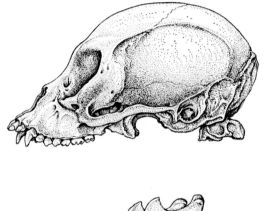

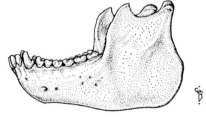

28 mm

FIG. 91. Skull and mandible of a Spider-monkey (*Ateles*).

ATELINAE (FIGS. 91, 92)

All three genera of this subfamily agree with *Alouatta* in the structure of the tail and its use as a tactile and investigating organ.

Ateles and *Brachyteles* have lost the pollex, the hands being elongated and used as hooks, locomotion being a modified form of brachiation. The Atelinae lack the extreme specialization of *Alouatta,* having skulls of normal shape. The woolly monkeys (*Lagothrix*) are sturdily built with dense woolly pelage and robust limbs. Spider monkeys contrast in their slender build, short trunks and elongated angular limbs. *Brachyteles* is intermediate in build and pelage. For a review of the Woolly Monkeys see Fooden (1963).

Fig. 92. Mexican Spider-monkey (*Ateles geoffroyi vellerosus*). Photo: courtesy of Dr. A. J. Sulzer, Communicable Diseases Center, Atlanta, Georgia.

20

Emergence of Old World Simians

Eocene and Oligocene Eurasian Simians

The earliest evidence available of the attainment of simian status by Primates in the Old World is derived from deposits of Oligocene age in Egypt – except for the problematical *Amphipithecus* and *Pondaungia* of the Upper Eocene of Burma.

The main peculiarity shown by *Amphipithecus*, of which only part of a mandible is known, is that the dental formula was the same as that of the New World monkeys. The creature was of large size and, if it proves to be a Primate rather than a condylarth ruminant, it may well have been on the line leading to the great apes. *Pondaungia*, from the same beds, has also been considered an early anthropomorph, but whereas the molars of *Amphipithecus* were elongated and 5-cusped, those of *Pondaungia* were quadrate and 4-cusped.

We are on surer ground when we consider the numerous fossils from the rich deposits of the Fayûm depression in Egypt. Here during the Oligocene the area was forested and traversed by sluggish streams and supported, besides small catarrhine monkeys, some larger forms that appear to have been precursors of the apes, as well as a wealth of other mammals, including rodents and the ancestors of hyraxes, sirenians and elephants.

Most generalized among the Fayûm Primates was the small *Parapithecus*, known principally from a mandible in which the two halves unite in V-fashion as in *Tarsius* and the Callithricidae and resembling the latter in size. The dental formula is usually given as $I_{\overline{2}}$; $C_{\overline{1}}$; $P_{\overline{2}}$; $M_{\overline{3}}$, i.e. the same as in all modern Old World Primates. However, there are some dubious points regarding the identification of the small canine; the original formula suggested having been $I_{\overline{1}}$; $C_{\overline{1}}$; $P_{\overline{3}}$; $M_{\overline{3}}$ and Simons (1970) after a re-study and comparison with other related Fayûm genera such as *Apidium*, has reverted to this formula, claiming that these genera, like platyrrhines, retained 3 pairs of premolars above and below. A third genus *Oligopithecus* was about the size of a Lion Marmoset, i.e. rather larger than *Parapithecus,* but differed in having a larger canine immediately in front of $P_{\overline{3}}$ and in several other points of dental morphology (see also Kälin (1961)).

Contemporary with the preceding were the supposedly gibbon-like *Propliopithecus* and *Aegyptopithecus,* which Simons (1965) declares to be undoubtedly referable to the dryopithecine group of anthropoid apes. The type mandible of *Moeripithecus,* moreover, is intermediate between these two.

It would, therefore, appear that two diverging evolutionary lines, one ancestral to the cynomorph monkeys and the other to the anthropoid apes, had already become differentiated during the Oligocene. These divergences became further established during the Miocene, when "the expansion of Anthropoidea to nearly their present limits of distribution seems to have been accomplished" (Simons, 1963).

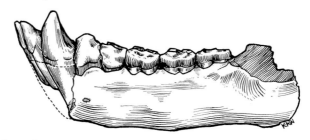

FIG. 93 *Pliopithecus antiquus.* Left side of the mandible with all the dental elements represented. An early fossil gibbon, dating from the Lower Pliocene of Europe but also recognized from Miocene strata elsewhere. After Hürzeler (1954) redrawn.

Miocene Eurasian and African Cynomorphs and Hominoids

In Eurasia fossil anthropoid apes (hominoids) have been recovered from Miocene rocks and deposits from Spain eastward through France, the Rhine Valley, Austria and thence to the Siwalik beds of India and possibly also from China. They are also well represented in the African Miocene. On the other hand, apart from some teeth from the Miocene of Kenya and some bony remnants discovered in association with hominoid fossils at Rusinga Island, Ombo and Maboko in Kenya, representatives of the catarrhine monkeys are distinctly scarce. This suggests that the cynomorph evolutionary radiation was largely post-Miocene in date, for there appear to be no climatic or ecological reasons for their lack of representation.

Among the Kenyan Miocene cynomorphs, pride of place must be given to *Victoriapithecus*, of which two species are recognized *V. macinnesi* and *V. leakeyi*, the former small and the latter larger. Upper molars are more quadrate in the latter, quadricuspidate and incipiently bilophodont, though some also retain remnants of a *crista obliqua* (connecting protocone and metacone). Relationship with earlier forms is closer to *Parapithecus* than to *Apidium* and not at all with the dryopithecine stem. An ulna referred to *V. leakeyi* that shows no proximal retroflexion seems to indicate lack of terrestrial adaptation (von Koenigswald, 1969).

Evidently in competition with the cynomorphs – possibly accounting for the paucity thereof – were the small gibbon-like apes of the genus *Limnopithecus* of East Africa and its contemporary European counterpart *Pliopithecus*, but later the reverse situation appears to have prevailed insofar as the small gibbon-like apes disappeared from Africa as a result of successful competition by arboreal tailed monkeys. Towards the end of the Miocene, for example, macaques had made their appearance in North Africa in the shape of *Macaca flandrini*, described by Arambourg (1959).

Pilbeam and Walker (1968) have broached the problem of the first indication of the dichotomy of the cynomorph evolutionary line into the two distinct branches, represented in the Recent primate fauna of both Africa and Asia, viz. the Cercopithecidae and the Colobidae. These two families are sundered by several contrasted morphological criteria which may be tabulated thus:

Cercopithecidae	Colobidae
Cheek pouches present	Absent
Stomach simple	Stomach complex, sacculated
Diet omnivorous	Diet specialized (leaves)
Molars with less marked transverse ridges	Molar transverse crests higher
$M_{\overline{3}}$ 4 or 5 cusped	$M_{\overline{3}}$ 5 cusped
Pollex well developed	Pollex feeble or absent

An upper molar of a small cercopithecid from the Miocene of Napak, Uganda, is described by Pilbeam and Walker (1968) as fully quadricuspid and without *crista obliqua,* while from the same deposits a frontal bone was found, which Walker believes to resemble more closely morphologically that of the Recent *Colobus.*

Contemporary with the gibbon-like species, the African Miocene also produced some larger hominoid apes, which contrast with the cynomorphs in their bunodont molars. These lack the sharp crests and oblong contours, being quadrate with 4 or 5 blunt rounded cusps (see Chapter 11, p. 85). *Proconsul* with 3 known species is the best known, *P. africanus* being represented by an almost complete skull, some limb bones and an almost entire hand skeleton. Though generalized anatomically, there are reasons for supposing that *Proconsul* species were in the ancestral line leading to the present day African apes, the chimpanzee and gorilla (see Le Gros Clark and Leakey (1951)). From a study of the fore limb and hand Napier and Davis (1959) inferred that *Proconsul* was probably semi-arboreal, a partial brachiator without unduly heavy commitments to brachiation.

An important group of hominoids known collectively as dryopithecines (from the typical genus *Dryopithecus*) made their earliest appearance in the Miocene of Eurasia, but these Miocene fossils provide little information of value in tracing the evolutionary history of the primate lineages, as they consist almost solely of mandibular fragments without skulls or limb bones. *Paidopithex,* known only from a femur from the late Miocene of Eppelsheim in Germany is of doubtful status; it may be dryopithecine or possibly pertain to *Oreopithecus.* Another taxon that includes limb bones is *Austriacopithecus* and seems to be dryopithecine, but it is uncertain which species is involved (see Thenius (1954) and Zapfe (1961)).

The earliest of this group to be discovered was *Dryopithecus fontani,* based on a mandible from the Miocene of Saint Gaudens, France (Lartet, 1856). This, combined with more recent recoveries, indicates that their owners were medium sized apes, comparable with a chimpanzee. They were provided with large canines, but smaller less spatulate incisors than their successors. Their lower molars show basically a 5-cusped biting surface and it was this feature that led Gregory (1922) to postulate that this group of apes could represent the common stock from which both man and the modern apes have been derived. Recent discoveries especially that of *Kenyapithecus* from the African Miocene have tended to confirm this early hypothesis.

Even more important in deciphering the origins of the human branch of the primate genealogical tree is the genus *Sivapithecus,* originally based on material of late Miocene and early Pliocene age from the Siwalik Hills at the foot of the Himalayas. The genus (or sub-genus) was later found in African Miocene deposits and reported by Le Gros Clark and Leakey (1951) as *S. africanus*, but Simons and Pilbeam (1965) have established that the African fossils are specifically identical with one of the Asiatic species, *S. sivalensis*. *Sivapithecus*, regarded by Simons and Pilbeam (1965) as no more than a subgenus of *Dryopithecus*, differs therefrom in having upper molars with moderately high, bunodont cusps, having their apices crowded towards the centre of the crown more than in typical *Dryopithecus* or *Proconsul*. A Carabelli cusp is often present (see Chapter 11, p. 82). The distribution of *Sivapithecus*, extended widely both to west and east of the type locality in the Siwaliks, west at least into Asia Minor and possibly into Europe (Spain); to the east it ranged into Yunnan (China).

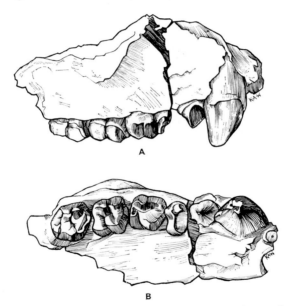

Fig. 94. *Sivapithecus indicus*, an early anthropoid dryopithecine dating from late Miocene until early Pliocene times. Damaged left maxilla, showing canine, damaged premolars and the molars. In A. the lateral aspects of the teeth are shown. In B. the teeth are shown in occlusion. Redrawn from Genet-Varcin after Pilgrim.

21

Primate Progress in the Eurasian and African Pliocene

The Pliocene is the period of efflorescence of the cynomorph monkeys, though the available evidence provides little guide to the immediate ancestry. The best preserved skull is that of *Libypithecus* from Middle or Late Pliocene beds in the Wadi Natrun, Egypt. It is remarkable for its prominent muzzle and high sagittal crest, thus foreshadowing the baboons, though also presenting affinities with macaques. Latest work (Jolly, 1967; Hill, 1970) has aligned it with the peculiar baboon-like Gelada (*Theropithecus*) of the Ethiopian highlands. Its brain cast, studied by Edinger (1938), shows nothing indicative of a status inferior to that of modern baboons.

Even more complete are the remains of *Mesopithecus* originally known from a complete skeleton from bone beds of Pontian age at Pikermi, near Athens, but now known from over a score of individuals from sites in Czechoslovakia, Rumania, Macedonia, Hungary, southern Russia and Iran, besides a few fragments of Miocene age recovered in Kenya, provisionally assigned to this genus. The name *Mesopithecus* was used to indicate its intermediate position between the colobid and the cercopithecid groups of present day monkeys, but the remains are generally considered to lean more towards the former. More important is the inference drawn by Patterson (1954) with respect to the ecology of *Mesopithecus*. This evidently differed from that of modern leaf-monkeys and suggests a less arboreal habitat since the associated fauna presents an open country or steppe facies. The limb bones have been studied comparatively by Gabis (1960) and by Napier (1970) who regards them are more macaque-like.

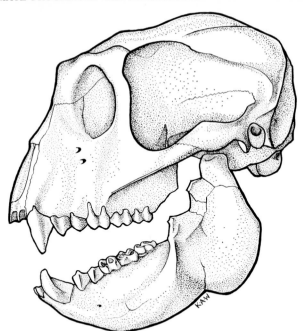

FIG. 95. Cranium of a fossil Old World monkey, *Mesopithecus pentelici*, from deposits of Pontian age (Lower Pliocene) in Greece, with features intermediate between the Cercopithecine and Colobine monkeys. Redrawn from Genet-Varcin after Gaudry.

Another extinct monkey that has been accorded colobid status is *Dolichopithecus* of which well preserved crania are known from deposits near Perpignan in southern France of slightly later age than those yielding *Mesopithecus*. It differs, however, from living colobids in its heavy almost baboon-like muzzle and shorter, more robust limbs but the short, broad nasal bones, expansion of the angular region of the mandible, small incisors and high cusped molars proclaim it definitely a colobid. Not so the second species, *D. arvernensis*, from the Villafranchian deposits at Senèze. This turns out to be a large, terrestrially adapted monkey, allied to the macaques (Jolly, 1967; Vogel, 1968), whereupon Simons (1970) suggests it should be assigned to the same taxon as certain Asiatic fossils typified by *Procynocephalus wimani*. Probably also assignable to *Procynocephalus* is *Paradolichopithecus* from Rumania with its large cercopithecid incisors. Jolly (1967) is convinced that *Procynocephalus* was an offshoot of the Asiatic Cercocebini, which took to terrestrial activities some time during the Pliocene, becoming extinct in the early Pleistocene.

Several species of typical macaques (*Macaca*) have been described on the basis of fragmentary material from Europe, Asia and
North Africa – presumably the modified descendents of the Miocene
M. flandrini. They include *M. pliocaena* from England and *M. prisca*
from Montpellier, France and also from sites in Germany and
Hungary, *M. libyca* from Wadi Natrun, Egypt, *M. sivalensis* from
northern India and two species (*M. anderssoni* and *M. robusta*)
from China which they did not reach until the end of the Pliocene
(von Koenigswald, 1969); Jolly (1967) thinks *Cynocephalus falconeri*
of the Siwaliks may be conspecific with *M. anderssoni*.

An important, though problematic, extinct type of the European
Pliocene is *Oreopithecus*, described originally in 1872 from the
Pontian lignites of Monte Bamboli in Tuscany, Italy. Much new
and complete material has been subsequently found and studied,
notably by Hürzeler (1949, 1954, 1958) and it is abundantly clear
that this was a tailless, erect-walking, bipedal primate making occa-
sional attempts at brachiation. Earliest students, before postcranial
bones were available, considered *Oreopithecus* a cercopithecid
allied to the macaques and baboons, but the general consensus
developed into the view that it occupied an intermediate position
between the cercopithecoids and the Hominoidea (e.g. Gregory
(1922)). Hürzeler assigns it to the human family (Hominidae), chiefly
on reasons derived from the dentition. However, though hominoid,
most authorities do not go so far as Hürzeler and the tendency is to
regard *Oreopithecus* as the representative of an independently
evolving sequence of an experimental nature, intermediate between
the great apes (Pongidae) and the Hominidae – and constituting
a family by itself (Oreopithecidae), thereby resuscitating a concept
advocated by Schwalbe (1916). Opposing views have been discussed by Straus (1963), Simpson (1963) and Simons (1963) (see also
Hill (1968)).

The Pliocene strata of the Siwalik Hills (the Salt Range in particular) have yielded abundant remains of primates chiefly, however,
pertaining to the anthropoid apes; for cynomorphs, though present,
are comparatively scarce. Cynomorphs representing such existing
genera as *Presbytis, Macaca* and *Papio* are known (Pilgrim, 1915).
Of the Pongidae there is ample evidence of the persistence of
Sivapithecus and other Dryopithecinae and many generic names
have been bestowed on individual fossils, probably unwarrantably.
Most important, however, is the genus *Ramapithecus*, based primarily on an upper jaw recovered in 1932 in the Simla Hills.
Originally classified with the Dryopithecinae, recent studies have

emphasized the dental resemblance to the Hominidae, in which family it is now included as representing a subfamily, Ramapithecinae, in an ancestral position on a level with later hominid types (Simons, 1961b).

22

Pleistocene and Holocene Cercopithecoidea

In general, mammalian genera during the Pleistocene have attained the same degree of evolutionary divergence as is still to be found today, albeit the representatives pertain to different species from the existing ones, usually differing, however, only in dimensions. In other words most modern genera are represented and few have become extinct but the (chronological) species are different, though, in some cases, they are directly ancestral to the modern congeners. Primates are no exception to this generalization. There are, however, changes in geographical distribution – brought about largely by the world wide climatic changes occurring during the successive glacial (or pluvial) phases and the intervening milder stadials. Thus the advance of the ice fields and resulting climatic deterioration affected Primates adversely, causing their extinction in Europe (with a single exception) and large areas of Asia.

The macaques (genus *Macaca*) are illustrative. Succeeding the late Miocene *M. flandrini* and Pliocene *M. prisca* and its allies, numerous bones have been unearthed throughout Europe and North Africa all pertaining to a species closely related to the existing Barbary Ape (*M. sylvanus*) of north-west Africa and Gibraltar (its last stronghold in Europe.) Thus we have *M. florentina* abundantly represented in Italy and other localities in Villafranchian deposits all over Europe as far north as Holland. *M. suevica* from caves in Württemberg, was associated with remains of *Rhinoceros mercki*; *M. majori*, a dwarf insular form from Sardinia and *M. tolosana* from near Toulouse, scarcely distinguishable from *M. sylvanus* – itself known from fossil and subfossil remains from the

Waalian interglacial at Episcopia and the Cromerian interglacial at Koneprusy and the Norfolk Forest Bed (Kurtén, 1968).

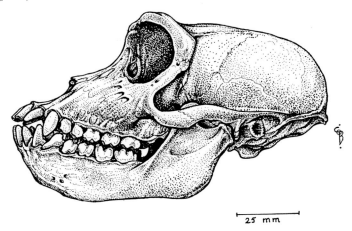

25 mm

FIG. 96. Cranium with mandible from the left side of a macaque—the Moor monkey (*Macaca maurus*). Note the high degree of prognathism and the dental formula characteristic of all Old World monkeys.

In Africa the Pleistocene has yielded remains of monkeys pertaining to lineages leading to the geladas (*Theropithecus*), the baboons (*Papio*) and a single example of a mangabey (*Cercocebus*). The genus *Simopithecus* was widely dispersed south of the Sahara; it was probably derived from *Libypithecus* and, in turn, was transformed into *Theropithecus*, with which *Simopithecus* shares the high crowned molars with deep valleys between the crests. Jolly (1970) is now inclined to make *Simopithecus* a synonym of *Theropithecus*; thus concurring in Freedman's (1957) belief that the Pleistocene genera of baboon-like fossils from South African sites (*Dinopithecus* and *Gorgopithecus*) are also related to *Theropithecus*.

As regards the savannah baboons, South African finds reveal the existence of forms less advanced in prognathism than the surviving *Papio*. They are represented by a diversity of species collectively assigned to the genus *Parapapio*, but there also appeared several true *Papio* species differing in various ways from the extant *P. ursinus* (Chacma baboon) of the area. So far, however, there are no known precursors of the forest dwelling baboons (*Mandrillus*) or of the other forest dwellers, such as *Cercopithecus*, that have produced a wide diversity of forms in the Recent fauna. This hiatus is probably to be explained by the nature of their habitat, which precludes the opportunity for fossilization of deceased individuals;

their remains are quickly disposed of by scavengers or destroyed by climatic vagaries. The mangabeys (*Cercocebus*) are, nevertheless, represented by a female mandible with cheek teeth *in situ* from the Middle Pleistocene of Kenya. The tooth-row exceeds in length by 45% those of the female of the species now represented in that region. South Africa has also yielded *Cercopithecoides*, now believed to be almost certainly a colobine of large size and so far the only evidence of the earlier occurrence of the African branch of the leaf monkeys, except for the recently described *Paracolobus*, based on a skull and nearly complete skeleton from the Baringo basin, central Kenya (Leakey, 1969).

Of Recent Old World monkeys (Cercopithecidae), the guenons of the genus *Cercopithecus* appear the most generalized. Their precise history is unknown and although they are primarily adapted to the tropical rain forest belt of Africa, a few inhabit the gallery forests along the principal rivers, whilst one group (*C. aethiops* complex) has advanced into the surrounding savannah to lead a partly terrestrial existence. The genus is consequently a large and extremely diverse and successful one. Well known examples are the elegant and colourful Diana monkey (*C. diana*) and the West African Green Monkey (*C. sabaeus*).

Offshoots of *Cercopithecus*, adapted for existence in swampy localities and sufficiently different morphologically to require separate generic status, are the small talapoins (*Miopithecus*) – the smallest Old World monkeys – and Allen's Swamp monkey (*Allenopithecus*). Both differ from *Cercopithecus* in having a catamenial swelling. Another offshoot of *Cercopithecus* and formerly classed therewith is the Red guenon or Dancing monkey (*Erythrocebus*) that has developed a harsh coat, long stilt-like limbs with short extremities; it is strongly adapted to semi-arid savannahs, but resorts to the trees for refuge or sleeping purposes.

The mangabeys (*Cercocebus*) much resemble guenons, but are larger, more powerful animals that have not evolved into such a wide variety of species as the guenons. They are, except for one gallery-forest inhabitant (*C.g. galeritus*) of the Tana River, confined to the rain forest, where they show a tendency to dichotomize into (a) a more arboreal group and (b) a terrestrially adapted group that subsists largely on the forest floor or in swamps. The arboreal group includes 2 species *C. albigena* and *C. aterrimus*, both of the Congo region; terrestrially adapted are the Sooty (*C. atys*) and White-collared Mangabeys (*C. torquatus*). The Golden-bellied *C. galeritus chrysogaster* is a swamp forest dweller. Problems relating

Fig. 97. Black-cheeked White-nosed Guenon (*Cercopithecus ascanius ascanius*) ♂ from Angola.

to the distribution of these and other African genera have been categorized and discussed by Tappen (1960).

Structurally similar to the Mangabeys, but even more terrestrially adapted are the macaques (*Macaca*), which are all Asiatic in distribution except *M. sylvanus*, the Barbary Ape, which is a relic from the Pleistocene still holding its own in North Africa and on the Rock of Gibraltar – its last outpost in Europe. Macaques are robust, stockily built monkeys varying in size from that of a small

G

poodle to a large mastiff. They are remarkable for a tendency to reduction of the tail. A long tail is retained in 3 species, the Common or Crab-eating macaque (*M. irus*) of Malaya and the Bonnet (*M. radiata*) and Toque monkeys (*M. sinica*) of South India and Ceylon respectively. In the Rhesus monkey (*M. mulatta*) of northern India the tail measures about half the body length, while in the Pig-tailed Macaque (*M. nemestrina*) of Burma and Malaya, it is shorter still as well as slender and carried in an arch. The organ is reduced to a mere tubercle in the Stump-tailed Macaque (*M. arctoides*) of South-east Asia and in the Celebesian Moor Macaque (*M. maurus*) and has disappeared altogether in *M. sylvanus*. Some macaques, such as *M. cyclopis* of Formosa, live in rocky terrain and many, like *M. irus*, are fond of water and swim readily. They are, on the whole,

FIG. 98. Stump-tailed Macaque (*Macaca arctoides*) adult female from Thailand. Courtesy of Dr. I. Bernstein, Yerkes Regional Primate Center.

a very successful group, adaptable to many environı
hardy, as witness their presence at high altitudes in the
(*M. assamensis*), China (*M. thibetana*) and Japan (*M. fusc*
they produce heavy coats against winter cold.

Macaques have evolved in northern Celebes a ı
baboon-like forms that comprise the genus (or subgenus) *Cynopithecus*; black, tailless or stump-tailed forms with crested heads
and heavy muzzles.

True baboons are purely African in range, where they are largely
addicted to savannah, parkland or rocky declivities even in the
Sahara, though here and there they enter the forest peripherally.
Five species are recognized, all large quadrupedal animals that con-
sort together in bands, dominated by a leader male. The Sacred
Baboon (*Papio hamadryas*) inhabits scrub-desert in north eastern
Africa and southern Arabia, ascending the hills in Abyssinia to 2000
metres altitude. It is remarkable for its sexual dimorphism, adult
males being twice the size of their consorts and distinguished from
them by their immense shoulder manes, grizzled grey in colour in
contrast to the drab olive pelage of the females. Other species are

(a) the heavily built *P. anubis*, which extends from west to east
    just south of the Sahara, with isolated populations in the hilly
    tracts of Jebel Marra, the Air massif and the Ennedi plateau;
(b) the more gracile Yellow Baboon (*P. cynocephalus*) of the
    eastern coastal tract;
(c) the large, dark, Chacma (*P. ursinus*) of South Africa and
(d) the smaller rusty coloured Guinea Baboon (*P. papio*) of the
    Guinea savannah.

Purely forest animals that feed on the forest floor in West Africa
are the stump-tailed Drill (*Mandrillus leucophaeus*) and Mandrill
(*M. sphinx*), both characterized by their elevated facial ridges and
vivid perineal colouration (see Hill (1970) and references there
cited).

The Gelada (*Theropithecus*), commonly but mistakenly classed
with baboons, is a quite different animal, rather a giant macaque but
with special features of its own (see above p. 130). It appears to be a
relic, now restricted to the high Abyssinian plateaux (above 2000m),
where it leads a rigorous existence, subsisting on grass, roots and
tubers unearthed manually. It is a derivitive of the much more
widely distributed Pleistocene genus *Simopithecus* (p. 181).

The Colobidae or leaf-eaters are today represented in Africa by
Colobus and in Asia by the langurs of which there are several

genera. *Colobus* comprises the black and black-and-white guerezas, which form the typical subgenus, the red or red-and-black forms (subgenus *Piliocolobus*) and the small, rather generalized Olive Colobus (subgenus *Procolobus*). *Procolobus* is confined to the Guinea rain forest, but the other forms range over the whole forest belt from west to east and also overflow into savannah or parkland, especially on the east, even into Abyssinia and the montane forest (see Rahm (1970)).

The principal genus of Asiatic leaf monkeys is *Presbytis*, a large and diverse group of graceful, lanky monkeys that occupy forest and scrub jungle, sometimes, as in the Indian *P. entellus*, feeding on the ground. Like *Colobus* the genus is susceptible to division into well marked subgenera. These are distinguished by morphological, including cranial, differences and also by the pelage of their newborn young (see Appendix A below). The genus ranges from Ceylon, north through the Indian peninsula into the Himalayas, thence eastwards through Assam, Burma, Thailand and the Malay Peninsula and is also represented in all the main Indonesian islands, as far east as Borneo, where leaf monkeys are richly represented.

Borneo is also the home of the specialized genus *Nasalis* – the Proboscis monkey – a form dwelling in riverside forests and addicted to swimming. Closely related are the snub-nosed monkeys of the genera *Rhinopithecus* and *Simias*, the former Chinese and the latter confined to the Mentawi islands off Sumatra. Close also to *Rhino-*

FIG. 99. Douc, (*Pygathrix nemaeus*) ♂. A representative of the Asiatic Leaf-monkeys from Indio China. × $\frac{10}{1}$. Photo W. C. Osman Hill.

pithecus are the Doucs (*Pygathrix*) of Vietnam and ⸓
noses, yellow faces, colourful pelage and relatively ⸓
a recent critical review of these rarer leaf-monkey
(1970), whose views are, however, somewhat revolutic
    Identification of the genera of Old World monkeys ⸓
fied by the aid of the subjoined key, taken from Hill (

Appendix A

KEY TO THE IDENTIFICATION OF THE GENERA OF RECENT OLD WORLD MONKEYS

Cheek-pouches present; stomach simple   I
Cheek-pouches lacking; stomach sacculated   II
  I. a. Cheek teeth without swollen bases; cusps not approximated;
         $M_{\overline{3}}$ without hypoconulid; orthognathous   2
  I. b. Cheek teeth with bases of crowns swollen; lingual and
         buccal pairs of cusps approximated; $M_{\overline{3}}$ with hypoconulid;
         prognathous   3
  2. a. Size moderate to large; no catamenial swelling in ♀
         (i) Size moderate; pelage not predominantly
             rufous   *Cercopithecus*
         (ii) Size larger (sexually dimorphic); pelage
              predominantly rufous   *Erythrocebus*
     b. A catamenial swelling in ♀
         (i) Size small; pelage pale green   *Miopithecus*
         (ii) Size moderate; pelage dark green   *Allenopithecus*
  3. a. Size moderate to large; body slender; tail very long   *Cercocebus*
     b. Size moderate to large; body robust; tail sometimes
        long, commonly reduced or even absent.
         (i) muzzle without swellings   *Macaca*
         (ii) muzzle with dorsal swellings; pelage black;
              head crested   *Cynopithecus*
     c. Size very large; body robust; muzzle greatly
        elongated; tail moderate or absent.
         (i) Muzzle with paired, raised, longitudinal
             swellings; tail a mere stump   *Mandrillus*
         (ii) Muzzle without swellings, merely ridges between
              dorsum and sides; tail longer and carried in an arch   *Papio*
         (iii) Muzzle shorter; rounded but ridged; tail longish,
               tufted at end; naked area on chest   *Theropithecus*
 II. a. Manus without pollex   *Colobus*
         (i) Pelage predominantly black   Subgenus *Colobus*
         (ii) Pelage predominantly reddish   Subgenus *Piliocolobus*
         (iii) Pelage olive   Subgenus *Procolobus*

B. Manus with pollex
- (i) External nose produced into a proboscis   *Nasalis*
- (ii) External nose retroussé
  - (a) Tail long   *Rhinopithecus*
  - (b) Tail short   *Simias*
- (iii) External nose normal; tail long
  - (a) Size moderate to large; tail longer than body; pelvic limbs longer than pectoral   *Presbytis*
  - (b) Size large; tail shorter than body; pectoral and pelvic limbs subequal   *Pygathrix*

23

Pleistocene and Holocene Hominoidea

The Pleistocene witnessed the geographical spread and steady and, in certain lineages, spectacular rise in evolutionary level of the Hominoidea including the emergence of man.

Gibbons (Hylobatidae)

Firstly must be considered the history of the gibbons (Hylobatidae) now confined to south-east Asia but, as has been seen, allegedly represented in the Oligocene and Miocene in Africa. However, as von Koenigswald (1968) has observed, the problem of whether the Hylobatidae ever occurred outside Asia has not been satisfactorily settled, insofar as all African and European fossils relegated to the ancestry of the gibbons have been disputed. Thus Ferembach (1958) has questioned the status of *Limnopithecus* while *Pliopithecus* of the European Middle Miocene, formerly regarded as a forerunner of *Hylobates* had a different skeletal structure (Zapfe, 1952) and a long tail (Ankel, 1965).

The oldest Asiatic gibbon-like fossil is the Upper Pliocene *Plio-pithecus posthumus* of Schlosser. Pleistocene gibbons (*Hylobates*) are known from sites in the Chinese provinces of Szechuan and Kwangsi and from Lower to Middle Pleistocene beds of central Java, while the larger siamang (*Symphalangus*) has also appeared in the highlands of southern Java, whence its remains have been recovered from rock-fissures (von Koenigswald, 1970).

Recent gibbons are regarded as constituting a separate family (Hylobatidae) though some reduce them to a subfamily (Hylo-

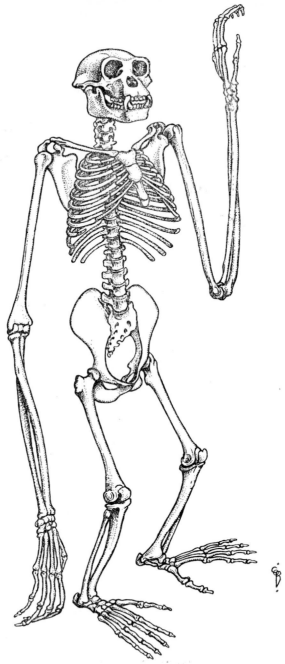

Fig. 100. Skeleton of a Siamang (*Symphalangus syndactylus*) an example of a modern gibbon. Redrawn from Schultz.

batinae) of the Pongidae, since they agree with the great apes in lacking a tail and in possessing a vermiform appendix to the caecum. On the other hand, they resemble catarrhine monkeys in possessing ischial callosities, so they clearly occupy an intermediate position. Their principal claim to attention is their perfection of the brachiating type of locomotion in the trees, combined with the power of erect bipedal progression on the ground (see Chapter 9, p. 66).

FIG. 101. Hoolock Gibbon (*Hylobates hoolock*). Adult female. Photo W. C. Osman Hill.

Two genera encompass all the existing gibbons; *Symphalangus*, for the siamang, is a large black form distinguished by the possession of an inflatable laryngeal air sac and *Hylobates*, which includes all the other species, the number of which is in dispute. Latest pronouncements are from Groves (1967b, 1968a, b, 1971) and Fooden (1970). Gibbons range geographically from Assam to China and through Indonesian islands to Borneo. Commonest and most variable is the Lar or White-handed Gibbon (*H. lar*) of the Malay Peninsula, Thailand and Indo-China. It is replaced in Sumatra by the dark-handed *H. agilis* and in Java by the Silvery Gibbon (*H. moloch*) which is represented by other races in Borneo. The Capped Gibbon (*H. pileatus*) appears distinct, though some consider it a subspecies of *H. lar*. Its range extends from Thailand to China. Differing structurally in several ways, the Concolor Gibbon (*H. concolor*), with six subspecies, from Hainan and Indo-China east of the Mekong, is made the basis of a separate subgenus, *Nomascus*, as is also the so-called Dwarf Siamang or Kloss's Gibbon, *H. (Brachytanites) klossi*, of the Mentawai islands.

Anthropoid Apes (Pongidae) and Mankind (Hominidae)

In tracing their history through the Pleistocene these two families of hominoids must be taken together for, during that period, there appeared so many forms that display a wide range of intermediacy in bodily structure, that it is well nigh impossible to come to a firm decision as to which of the two families any particular genus pertains. (see, however, Hürzeler (1968)).

Though *Dryopithecus* itself did not survive into the Pleistocene, the genus *Gigantopithecus*, apparently derived from it, did. This was a gigantic ape-like creature whose dental remains have been recovered from southern China. Like the dryopithecines and unlike modern apes, it had relatively small vertically planted incisors, a weak simian shelf and huge molars. The mandible was deeper and more robust than in *Dryopithecus, Proconsul* or *Gorilla*. Simons and Pilbeam (1965) consider *Gigantopithecus* the aberrant endproduct of a stock which, during the Miocene or Pliocene, split off from the main dryopithecine assemblage. The mandible and dentition suggest a vegetarian diet comparable to that of the modern gorilla.

More important Pleistocene hominoids are those derived mainly from South and East Africa collectively referred to as the australopithecines, as they constitute the subfamily Australopithecinae or southern ape-men. Whether this group is assignable to the Pongidae or the Hominidae is debatable, but that is purely a question of semantics (see Tobias (1968)). There can be no doubt that they fill an important gap in the phylogenesis of the later expressions of evolution to the human level.

Since the original discovery at Taung of a juvenile skull (with a complete natural endocast) named *Australopithecus africanus* by Dart (1925), many additional specimens of this or allied creatures have been recovered. There was an unfortunate tendency to label each new discovery with a new name and a consequent burdening of the literature with invalid taxa, e.g. generic terms like *Plesianthropus, Paranthropus, Zinjanthropus* and *Paraustralopithecus*. An overall consideration of the group leads to the conclusion that, although considerable evolutionary adaptation occurred within the group, taxonomically all can conveniently be regarded as representatives of a single genus *Australopithecus*, with no more than 3 species, at present known, *A. africanus*, *A. robustus* and *A. boisei*. Tobias (1968) considers, therefore, that the subfamily is unnecessary and the genus should be included, like *Ramapithecus* (p. 178), within the Hominidae – a view advocated by Dart in 1948.

Australopithecus manifestly approaches closely the human level. Its brain is enlarged in those areas (association areas) between the centres for vision, hearing and general sensibility (especially the parietal association area); its dentition is more human than pongid (small canines, bicuspid premolars) and it had assumed an erect, bipedal gait. The erect attitude perhaps was not quite finally perfected and the brain still had a long period of improvement ahead. Nevertheless the hands were released for the wielding of weapons and there is some evidence of tool using and tool-making – at least with the gracile *A. africanus*, who may have preyed upon its larger vegetarian contemporaries.

Until the time of the discovery of australopithecines and other hominids at Olduvai in East Africa, only two South African forms, *A. africanus* and *A. robustus* – i.e. a Lower Pleistocene gracile and a Middle Pleistocene robust form – were known and their phylogenetic relations with each other and with *Homo* (whose remains were contemporaneus with *A. robustus* in the Middle Pleistocene) were susceptible to various interpretations. With the Olduvai discoveries, however, the picture changed, for these included the Lower Pleistocene hyperrobust *A. boisei*, as well as an earlier *Homo*, namely *H. habilis*.

Tobias (1968) has adduced evidence to show that *A. africanus* is close to the line of hominization; it is also near in structure to the common ancestor of the whole australopithecine group. This is supported by the dental and mandibular similarities to *Ramapithecus* (as exemplified by its African representative "*Kenyapithecus*"). On the other hand, the large toothed *A. robustus* and *A. boisei* are to be regarded as specialized vegetable feeders, splitting off the main evolutionary line leading to man, just as we saw with *Giganto-*

pithecus in its phyletic relationship with the Dryopithecinae. The Pleistocene genealogical tree may therefore be expressed thus:

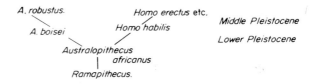

A. robustus.
A. boisei
Homo erectus etc.
Homo habilis
Australopithecus
    africanus
Ramapithecus.
Middle Pleistocene
Lower Pleistocene

During the Pleistocene, the human lineage gave rise, after *H. habilis,* to the forms earlier listed under the generic name *Pithecanthropus,* first known from Dubois' finds (1898) in Java, but since uncovered in China, North and East Africa and elsewhere. They are characterized by their flattened cranial vaults, a nuchal plane forming an angle with the vault and heavy supraorbital ridges; but their brains were considerably larger than in any of the australopithecines. Modern writers place them in the genus *Homo* as *H. erectus,* but in speaking of the evolutionary level which they had attained, the term "pithecanthropines" is still used for them collectively. In the later Pleistocene the pithecanthropines had given rise to two divergent evolutionary lines, one leading to Neanderthal Man (*Homo neanderthalensis*) the other to a modern type (*Homo sapiens*) that survived into the Holocene, whereas Neanderthal Man became extinct before the end of the Pleistocene without leaving any descendants (see Howell (1967) and works there cited).

As regards the Pleistocene representatives of the modern great apes the only documents of their existence are four teeth recovered from Kenya and reported by Lönnberg (1937) under the name *Proconsuloides naivashae.* These appear to have pertained to a chimpanzee and were so called from a likeness to *Proconsul,* from which their owners were presumably descended. Their distinctiveness from the teeth of modern chimpanzees is insufficient, in the opinion of Hopwood and Hollyfield (1954), to sever them from the genus *Pan.*

Holocene great apes fall naturally into the three genera *Pongo* orang-utans), *Pan* (chimpanzees) and *Gorilla.* The last two mentioned are confined to the African rain forest, the orangs being Asiatic, occurring today only in Sumatra and Borneo. The two African forms are more closely related to one another than either is to *Pongo* and on this account some would place them together in the genus *Pan.*

Orangs must at some time or other have occurred on the Asiatic mainland; their presence in Sumatra and Borneo proclaims them as relict populations of a once more widely dispersed species. They

FIG. 102. Orang-utan (*Pongo pygmaeus abeli*), adult male. Courtesy of Mr S. H. Benson from Hill (1954).

are extremely arboreal and adapted for life in the rain forests, especially along rivers and swamps. Their diet is vegetarian, chiefly comprized of fruits (durian, rambutan), leaves, bark and occasionally eggs. Orangs display marked sexual dimorphism, males weighing twice as much as females, from whom they also differ in several secondary sexual characters. Sumatran orangs tend to exceed the Bornean animals in size and are also paler in coat colour, which is some shade of fiery red, tending rather to sandy or orange in Sumatran and to maroon in Bornean examples. The face is bare and of a blue-black or slaty hue, except for an orange beard and moustache in old males. Old males also develop oval fibro-fatty cheek pads on either side and a dependent transverse dewlap that is inflatable from a laryngeal air-sac. Ears are small, but resemble human ears in morphology. The head is high and domed, without heavy brow ridges and the face concave, with prognathous jaws. Arms are very long, reaching the ankles when the animal stands erect. The hands are long, with curved digits, capable of combined use as a hook in body suspension; the pollex being relatively short. Legs are relatively short, provided with long, narrow feet held, when at rest, in an inverted and flexed position, with long, curved toes and short, sometimes nailless hallux.

The dentition is of the typical pongid pattern, as in *Pan*, but the enamel of the molars shows considerable secondary wrinkling. Supernumerary molars are frequent. Ribs 12 pairs; carpus with an os centrale.

Chimpanzees (*Pan*) occupy the Guinean and Congo rain-forests and extend into deciduous forest and forest savannah in Guinea and to the east of the Congo rain-forest in Uganda and Tanganyika. Though partially arboreal, they are less so than orangs, often being found on the ground but retiring to the trees for sleeping, where they construct sleeping platforms by weaving together neighbouring leafy twigs. The diet is primarily vegetarian (fruit, nuts, bark, leaves and seeds) but supplemented by insects (especially termites and ants) and fish. For recent field studies, consult Stanley (1919), Nissen (1931), Goodall (1962, 1963, 1965, 1967), Kortlandt (1962, 1966, 1968), Reynolds and Reynolds (1955) and Reynolds (1965).

Chimpanzees are slighter than orangs or gorillas; adults standing erect reach a height of 1·5m (compared with 1·37 for *Pongo* and 1·8m for *Gorilla*). There is less sexual dimorphism in size. They differ from the other apes and approach *Homo* in their smaller canines and the fact that $M_{\overline{3}}$ is smaller than $M_{\overline{2}}$. The pelage is black, with some white hairs on the chin and in juveniles around the anus; greying occurs on the back of elderly specimens. Baldness

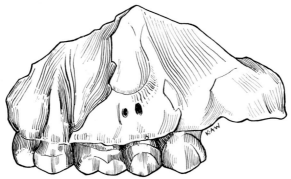

FIG. 103. *Ramapithecus brevirostris*; a Pliocene hominoid primate from the Siwaliks of Northern India.   Maxillary fragments with $P^3 - M^3$ in situ. $\times 1 \cdot 5$. Redrawn from Simons (1961b).

frequently develops as far back as the parietal region in both sexes. The facial skin may be pale flesh-colour, bronzed, blotched or uniformly black, differing with age and racially. Elsewhere, e.g. ears, similar changes occur. The ventral surface is sparcely haired and pigmented skin shows through. Ears are large and prominent. Lips are mobile and capable of protrusion. In the skull heavy brow ridges are developed, but the brain case is rounded and smooth; a small sagittal crest occasionally develops but not a lambdoid crest. The mandible is less massive than in *Pongo;* the face, however, is prognathous. Ribs 13 pairs; carpus with os centrale fused to scaphoid. Arms are longer than legs, but the discrepancy is less than in *Pongo,* the legs being sturdier. A laryngeal air sac is present, but less extensive than in *Pongo* and never forming a dewlap.

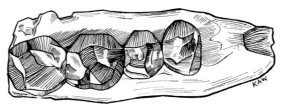

FIG. 104. *Ramapithecus brevirostris*. The two lower premolars and the foremost of the molars. $\times 1 \cdot 5$. Redrawn from Simons (1961b).

Besides the common chimpanzee (*P. troglodytes*) with its geographical variaents, there exists, in the forests south of the Congo, a pygmy form (*P. paniscus*). This differs not only in its smaller size (stature $1 \cdot 01$m) but also in its more gracile (paedomorphic) build and limb proportions, producing a more gibbon-like aspect. The face is black from birth, with a tendency to flesh-coloured labial margins; the ears are small and black and the nose broad and padded resembling a gorilla's. The second and third toes tend to exhibit syndactyly. The species shows temperamental differences from ordinary chimpanzees, a more horizontal stance in quadrupedal progression and a markedly different voice.

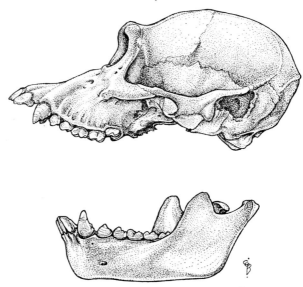

FIG. 105. Cranium and mandible of an adult female chimpanzee (*Pan troglodytes*).$\frac{2}{1}$

Gorillas (*Gorilla*) are more restricted in range than chimpanzees, their distribution being discontinuous. The Lowland Gorilla (*G. gorilla*) occurs in the west from about the Cross River in south eastern Nigeria through Cameroon, Gabon and the former French Congo south to the Congo River. It reappears in the east as the Mountain Gorilla (*G. beringei*) in the Virunga volcanoes and other high mountains to the north and east of Lake Kivu. Scattered populations occurring in the lowlands west of the Upper Congo (River Lualaba) have been assigned to a subspecies of *G. gorilla* as *G.g. manyema* (Groves, 1967a).

Despite their huge size and powerful build (stature 1·8m) gorillas are peaceful, inoffensive and purely vegetarian-feeding inhabitants of the rain-forest. They spend 90% of their time on the ground, foraging and moving about quadrupedally. Only the youngsters resort regularly to the trees. Adults build sleeping platforms on or near the ground.

There is some sexual dimorphism in size, males weighing twice as much as females. The arm span exceeds the stature, reaching 2·3m. The chief characters are the bare, black face surmounted by a broad flat nose, flanked by padded alar folds; ears are small and black. The pelage is dull black to iron-grey, with a tendency to rusty on the crown. A conical fibro-fatty pad develops on the crown of adult males, producing a helmet-like effect. This is enhanced by the high sagittal and nuchal crests on the skull. A bull-necked effect is produced by the great height of the spinous processes of the cervical vertebrae, flanked by the powerful nuchal muscles. The hands are very broad, with short, thick digits, but the pollex is relatively shorter than in man and more feebly muscled. The foot has a long heel and is applied plantigrade. The hallux is short, stout and well abducted in G. gorilla, longer, but more in alignment with the other toes in G. beringei.

In the skeleton there are 13 pairs of ribs and the os centrale is fused with the scaphoid, as in *Pan*. $M_{\overline{1}}<M_{\overline{2}}<M_{\overline{3}}$. Cusps of molars more sharply defined than in *Pan*.

Field studies on the gorilla, with special reference to their social behaviour, have been reported by Akeley (1922), Maxwell (1928), Coolidge (1929), Bingham (1932), Pitman (1935), Donisthorpe (1958), Kawai and Mizuhara (1959), Osborn (1963), Sabater Pi (1964), Emlen and Schaller (1960) and Schaller (1963, 1965).

Identification of the genera of Recent Hominoidea will be facilitated by the following key:

Appendix B

KEY TO THE IDENTIFICATION OF THE GENERA OF RECENT ANTHROPOID APES

Size small to moderate; arms greatly elongated and slender; pollex separated from other digits down to base of metacarpal; ischial callosities present   I

Size large, arms elongated to lesser degree and very robust; pollex short and not separated beyond proximal phalanx   II

  I. a. A laryngeal sac present, covered with naked integument; pedal digits II and III syndactylous   *Symphalang*

     b. No laryngeal sac marked externally by naked skin; pedal digits II and III not syndactylous   *Hyl*

II. a. Ears small, pelage red; males with cheek excrescences

   b. Ears small, pelage grey-black; males with
      excrescences on vertex; hands short and broad,
      penis short and thick   *Gorilla*
   c. Ears large, outstanding; pelage glossy black; males
      without facial or cranial excrescences; hands long
      and slender; penis long, narrowing to a fine apex   *Pan*

    The post-Pleistocene history of *Homo* is outside the scope of this
work. To do it justice would involve doubling the size of this book.
In any event the ground is covered in a number of treatises that are
readily accessible to the student. The paper by Howell (1967) already quoted will serve as a resumé of recent work, while an overall picture of human history is provided by Darlington (1969).

References

Abel, O. (1931). "Die Stellung des Menschen im Rahmen der Wirbeltiere". G. Fischer, Jena.

Akeley, C. E. (1922). Hunting gorillas in Central Africa. *Wld's Work.* **44**, 169–183.

Amoroso, E. C. (1961). Placentation. *In* "Marshall's Physiology of Reproduction" (A. S. Parkes, ed.) Vol. 2. Little, Brown and Company, Boston.

Anderson, J. (1878). "Anatomical and Zoological Researches: Comprising an Account of the Zoological Results of the Two Expeditions to Western Yunnan in 1868 and 1875" B. Quarich, London.

Angst, R. (1970). Über die Schädelkämme der Primaten. *Natur Mus.* **100**, 293–302.

Ankel, F. (1965). Der canalis sacralis als Indikator für die Länge der Caudalregion der Primaten. *Folia primat.* **3**, 263–276.

Ansell, W. F. H. (1963). Additional breeding data on Northern Rhodesian mammals, *Puku*, **1**, 9–28.

Arambourg, C. (1959). Vertébrés continentaux du Miocene supérieure de l'Afrique du Nord. *Publs. Sen. Geol. Algerié. Pal.* **4**, 1–59.

Ashton, E. H. and Oxnard, C. E. (1963). The musculature of the primate shoulder. *Trans. zool. Soc. Lond.*, **29**, 553–650.

Ashton, E. H. and Oxnard, C. E. (1964a). Functional adaptations in the primate shoulder. *Proc. zool. Soc. Lond.* **142**, 49–66.

Ashton, E. H. and Oxnard, C. E. (1964b). Locomotor patterns in primates. *Proc. zool. Soc. Lond.* **142**, 1–28.

Aubert, E. (1929). Recherches anatomiques sur les sinus osseux des simiens. *C. r. Ass. Anat.* **24**, 573–575.

Ayer, A. A. (1948). "The Anatomy of *Semnopithecus entellus*". Indian Publ. House Ltd., Madras.

Bartels, P. (1905). Über die Nebenräume der Kehlkopfhohle. *Z. Morph. Anthrop.* **8**, 11–61.

Bingham, H. C. (1932). Gorillas in a native habitat. *Carnegie Inst. Wash. Publ.* **426**, 1–66.

de Blainville, H. M. D. (1838). *L'Institut, Paris* **6**, (243), 275. (*Amphitherium*).

Bluntschli, H. (1938). Die Sublingua und Lyssa der Lemuriden-zunge. *Biomorphosis*, **1**, 127–149.

Bolk, L. (1907). Beitrage zur Affenanatomie, VI, Zur Entwicklung und vergleichenden Anatomie des Tractus urethro-vaginalis der Primaten. *Z. Morph. Anthrop.* **10**, 250–316.

Broom, R. (1915). On the organ of Jacobson and its relations in the Insectivora. *Proc. zool. Soc. Lond.* 157–162, 347–354.

Buettner-Janusch, J. (1963). An introduction to the primates. *In* "Evolutionary and Genetic Biology of Primates" (J. Buettner-Janusch, ed.) Academic Press, New York and London.

Buffon, G. L. L., Comte de (1765). "Histoire Naturelle, Générale et Particulière". l'Imprimerie Royale, Paris.

Butler, H. (1957). The breeding cycle of the Senegal galago *Galago senegalensis senegalensis*. *Proc. zool. Soc. Lond.* **129**, 147–149.

Butler, H. (1960). Some notes on the breeding cycle of the Senegal galago *Galago senegalensis senegalensis* in the Sudan. *Proc. zool. Soc. Lond.* **135**, 423–430.

Butler, H. (1964). The reproductive biology of a strepsirrhine (*Galago senegalensis senegalensis*). *Int. Rev. gen. exp. Zool.* **1**, 241–246.

Butler, H. (1966a). Observations on the menstrual cycle of the grivet monkey (*Cercopithecus aethiops*) in the Sudan. *Folia primat.* **4**, 194–205.

Butler, H. (1966b). Seasonal breeding of the Senegal galago (*Galago senegalensis senegalensis*) in the Nuba mountains. *Folia primat.* **5**, 165–175.

Butler, H. (1967). The oestrous cycle of the Senegal bush baby (*Galago senegalensis senegalensis*) in the Sudan. *J. Zool. Lond.* **151**, 143–162.

Butler, P. M. (1956). The skull of *Ictops* and the classification of the Insectivora. *Proc. zool. Soc. Lond.* **126**, 453–481.

Cabrera, A. (1925). "Genera Mammalium: Insectivora Galeopithecia" 232 pp. Museo Nacional de Ciencias Naturales, Madrid.

Cantor, T. (1846). Catalogue of Mammalia inhabiting the Malayan Peninsula and Islands. *J. Asiat. Soc. Beng.* **15**, 188–190.

Carlsson, A. (1909). Die Macroscelididae und ihre Beziehungen zu den übrigen Insectivoren. *Zool. Jb. (Syst).* **28**, 349–400.

Carlsson, A. (1922). Über die Tupaiidae und ihre Beziehungen zu den Insectivora und den Prosimiae. *Acta. zool. Stockh.* **3**, 227–270.

Carpenter, C. R. (1940). A field study in Siam of the behaviour and social relations of the gibbon (*Hylobates lar*). *Comp. Psychol. Monogr.* **16**, (5), 1–122.

Carpenter, C. R. (1941). The menstrual cycle and body temperature in two gibbons. *Anat. Rec.* **79**, 291–296.

Catchpole, H. R. and Fulton, J. F. (1943). The oestrous cycle of *Tarsius*. *J. Mammal.* **24**, 90–93.

Cave, A. J. E. and Haines, R. W. (1940). The paranasal sinuses of the anthropoid apes. *J. Anat.* **74**, 493–523; 558.

Clark, J. (1941). An anaptomorphid primate from the Oligocene of Montana. *J. Paleont.* **15**, 562–563.

Clark, W. E. Le Gros (1926). On the anatomy of the pen-tailed tree-shrew (*Ptilocercus lowii*). *Proc. zool. Soc. Lond.* 1179–1309.

Clark, W. E. Le Gros (1930). The thalamus of *Tarsius. J. Anat.* **64**, (3), 371–414.

Clark, W. E. Le Gros (1931). The brain of *Microcebus murinus. Proc. zool. Soc. Lond.* 463–486.

Clark, W. E. Le Gros (1932). A morphological study of the lateral geniculate body. *Br. J. Ophthal.* **16**, 264–284.

Clark, W. E. Le Gros (1933). The brain of the Insectivora. *Proc. zool. Soc. Lond.* 975–1013.

Clark, W. E. Le Gros (1934). "Early Forerunners of Man" Baillière, Tindall and Cox, London.

Clark, W. E. Le Gros (1959). "The Antecedents of Man" Edinburgh University Press.

Clark, W. E. Le Gros and Leakey, L. S. B. (1951). The Miocene Hominoidea of East Africa. *In* "Fossil Mammals of Africa" No. 1, 1–117. British Museum (Natural History), London.

Clark, W. E. Le Gros and Thomas, D. P. (1952). A miocene lemuroid skull from East Africa. *In* "Fossil Mammals of Africa" No. 9, 6 pp. British Museum (Natural History), London.

Coimbra-Filho, A. F. (1969). Mico-Leao, *Leontideus rosalia* (Linnaeus, 1766) situaçao atual do espécie no Brasil. *An Acad. bras. Ciens.* **41**, suppl. 29–52.

Coimbra-Filho, A. F. (1970a). Consideraçoes gerais e situação atual dos micos-leoes, *Leontideus chrysomelas* (Kuhl, 1870), e *Leontideus chrysopygus* (Mikan, 1873). *Revta. bras. Biol.* **30**, 249–268.

Coimbra-Filho, A. F. (1970b). Acerea da redesioberta de *Leontideus chrysopygus* (Mikan, 1823) e apontamentos sóbre sua ecologia. *Revta. bras. Biol.* **30**, 609–615.

Conaway, C. H., and Sorenson, M. W. (1965). Reproduction in tree shrews. *J. Reprod. Fertil,* **9**, 389–390.

Coolidge, H. (1929). A revision of the genus *Gorilla. Mem. Mus. comp. Zool. Harv.* **50**, 291–381.

Corner, G. W. (1923). Ovulation and menstruation in *Macacus rhesus. Contrib. Embryol.* **15**, 73–101.

Cowgill, U. M., Bishop, A., Andrew R. J. and Hutchinson, G. E. (1962). An apparent lunar periodicity in the sexual cycle of certain prosimians. *Proc. natn. Acad. Sci. U.S.A.* **48**, 238–241.

Darlington, C. D. (1969). "The Evolution of Man and Society" Allen and Unwin, London.

Dart, R. A. (1925). *Australopithecus africanus:* the man-ape of South Africa. *Nature, Lond.* **115**, 195–199.

H

Dart, R. A. (1948). The Makapansgat proto-human *Australopithecus prometheus. Am. J. phys. Anthrop*, **6**, 259–284.

Dart, R. A. (1963). The carnivorous propensity of baboons. *Symp. zool. Soc. Lond.* **10**, 49–56.

Day, M. H. and Walker, A. C. (1969). New prosimian remains from early Tertiary deposits of southern England. *Folia primat.* **10**, 139–145.

Dempsey, E. W. (1939). The reproductive cycle of New World monkeys. *Am. J. Anat.*, **64**, 381–405.

DeVore, I. and Hall, K. R. L. (1965). Baboon ecology. *In* "Primate Behaviour" (I. DeVore, ed.) Holt, Rinehart and Winston, New York.

DeVore, I. and Washburn, S. L. (1963). Baboon ecology and human evolution. *In* "African Ecology and Human Evolution" (F. C. Howell and F. Bourlière, eds.) Wenner-Gren Foundation, New York.

Donisthorpe, J. (1958). A pilot study of the Mountain Gorilla, (*G. g. beringei*) in S.W. Uganda, February to September, 1957. *S. Afri. J. Sci.* **54**, 195–217.

Doyle, G. A., Pelletier, A. and Bekker, T. (1967). Courtship, mating and parturition in the lesser bushbaby (*Galago senegalensis moholi*) under semi-natural conditions. *Folia primat.* **7**, 169–197.

Eckstein, P. (1958). Internal reproductive organs. *In* "Primatologia" (H. Hofer, A. H. Schultz and D. Starck, eds.) Vol. 3, part 1, 542–629. Karger, Basel.

Eckstein, P. and Zuckerman, S. (1962). Morphology of the reproductive tract. *In* "Marshall's Physiology of Reproduction" (A. S. Parkes, ed.) 3rd edition, Vol. 1, part 1, 43–155. Little, Brown and Company, Boston. Oestrous cycle in the mammalia. *Idem* 226–396.

Edinger, T. (1938). Mitteilungen über Wirbeltierreste aus dem Mittel-pliocän der Natrontales (Ägypten) 9. Das Gehirn von *Libypithecus. Zentbl. Miner. Geol. Paläont.* 122–128.

Elliott Smith, G. (1927). "Essays on the Evolution of Man" 2nd. edition. Oxford University Press.

Emlen, J. T. and Schaller, G. B. (1960). Distribution and status of the mountain gorilla. *Zoologica, N.Y.*, **45**, 41–52.

Evans, F. G. (1942). The osteology and relationships of the elephant-shrews (Macroscelididae). *Bull. Am. Mus. nat. Hist.* **80**, 85–125.

Fahrenholz, C. (1937). Drüsen der Mundhohle. *In* "Handbuch der Vergliechende Anatomie der Wirbeltiere" (L. Bolk, E. Göppert, E. W. H. Kallius and W. Lubosch, eds.) Vol. 3. Urban and Schwarzenburg, Berlin and Vienna.

Ferembach, D. (1958). Les Limnopitheques de Kenya. *Ann. Paleont.* **44**, 151–249.

Flower, S. S. (1933). Breeding season of lemurs. *Proc. zool. Soc. Lond.* **1**, 317.

Flower, W. H. and Lydekker, R. (1891). "An Introduction to the Study of Mammals, Living and Extinct" A. and C. Black, London.

Fooden, J. (1963). A revision of the Woolly monkeys (Genus *Lagothrix*) *J. Mammal.* **44**, 213–247.

Fooden, J. (1964). Stomach contents and gastro-intestinal proportions in wild-shot Guianian monkeys. *Am. J. phys. Anthrop.* **22**, 227–232.

Fooden, J. (1970). Color phase in gibbons. *Evolution.* **23**, 627–644.

Frechkop, S. (1949). Le crâne de l'homme en tant que crâne de mammifere (1). *Bull Inst. R. Sci. nat. Belg.* **25**, 12pp.

Freedman, L. (1957). The fossil Cercopithecoidea of South Africa. *Ann. Transv. Mus.* **23**, 121–262.

Gabis, R. V. (1960). Les os des membres des singes cynomorphes. *Mammalia,* **24**, 577–607.

Gautier, J. P. and Gautier-Hion, A. (1969). Les associations polyspécifiques chez les Cercopithecidae du Gabon. *Terre Vie.* 164–201.

Gautier-Hion, A. (1968). Étude du cycle annuel de réproduction du talapoin (*Miopithecus talapoin*) vivant dans son milieu natural. *Biol. gabonica* **4**, (2), 163–173.

Gazin, C. L. (1958). A review of the Middle and Upper Eocene primates of North America. *Smithson. misc. Collns.* **126**, 1–112.

Gensch, W. (1963). Breeding Tupaias, *Tupaia glis,* at Dresden Zoo. *In* Jarvis and Morris. *Int. Zoo. Yb* **4**, 75–76.

Gérard, P. (1932). Études sur l'ovogenèse et l'ontogenèse chez les lemuriens du genre *Galago. Archs. Biol., Paris* **43**, 93–151.

Gill, T. (1872). Arrangement of the main families of Mammals with analytical tables. *Smithson. Misc. Coll.* **11**, i–vi; 1–98.

Glasstone, S. (1938). A comparative study of the development *in vivo* and *in vitro* of rat and rabbit molars. *Proc. R. Soc. (B),* **126**, 315.

Goodall, J. (1962). Nest building behaviour in the free ranging chimpanzee. *Ann. N.Y. Acad. Sci.* **102**, 455–467.

Goodall, J. (1963). Feeding behaviour of wild chimpanzees. *Symp. zool. Soc. Lond.* **10**, 39–47.

Goodall, J. (1965). Chimpanzees of the Gombe Stream Reserve. *In* "Primate Behaviour" (I. DeVore, ed.) Holt, Rinehart and Winston, New York.

Goodall, J. (1967). "My Friends the Wild Chimpanzees". National Geographic Society, Washington, D.C.

Goodman, M. (1963). Man's place in the phylogeny of the primates as reflected in serum proteins. *In* "Classification and Human Evolution" (S. L. Washburn, ed.) 204–234. Aldine Publishing Company, Chicago.

Göppert, E. (1931). Atmungswege. *In* "Handbuch der Vergliechende Anatomie der Wirbeltiere" (L. Bolk, E. Göppert, E. W. H. Kallius and W. Lubosch, eds.) Vol. 3. Urban and Schwarzenburg, Berlin and Vienna.

Grand, T. L. and Lorenz, R. (1968). Functional analysis of the hip joint in *Tarsius. Folia primat.* **9**, 161–181.

Grandidier, G. (1904). Un nouveau lémurien fossile de France, le *Pronycticebus gaudryi. Bull. Mus. Hist. nat. Paris,* **10**, 9–13.

Grassé, P. (1955). Mammifères in *Traité de Zoologie,* vol. **17**, Paris: Masson pp. 1–2300.

Gregory, W. K. (1910). The orders of mammals. *Bull. Am. Mus. nat. Hist.*
   27, 1–54.
Gregory, W. K. (1913). Relationship of the Tupaiidae and of Eocene
   lemurs, especially *Notharctus. Bull. geol. Soc. Am.* **24**, 247–252.
Gregory, W. K. (1920a). On the structure and relations of *Notharctus*, an
   American Eocene primate; *Am. Mus. nat. Hist. Mem.* **3**, 49–243.
Gregory, W. K. (1920b). The origin and evolution of the human dentition.
   A palaeontological review. *J. dent. Res.* **2**, 89–183, 215–283, 357–426,
   607–711.
Gregory, W. K. (1921). The origin and evolution of the human dentition.
   A palaontological review. *J. dent. Res.* **3**, 87–228.
Gregory, W. K. (1922). "The Origin and Evolution of the Human
   Dentition" Williams and Wilkins, Baltimore.
Grosser, O. (1933). Human and comparative placentation. *Lancet,* **224**,
   999–1001; 1053–1058.
Groves, C. P. (1967a). Ecology and taxonomy of the gorilla. *Nature, Lond.*
   213, 890–893.
Groves, C. P. (1967b). Geographic variation in the hoolock or white-
   browed gibbon (*Hylobates hoolock* Harlan, 1834) *Folia primat.* **7**,
   276–283.
Groves, C. P. (1968a). The classification of the gibbons (Primates, Pongi-
   dae). *Z. Säugetierk.* **33**, 239–246.
Groves C. P. (1968b). A new subspecies of white-handed gibbon from
   northern Thailand *Hylobates lar carpenteri. Proc. biol. Soc. Wash.*
   81, 625–628.
Groves, C. P. (1970). The forgotten leaf-eaters, and the phylogeny of the
   Colobinae. *In* "Old World Monkeys: Evolution, Systematics and
   Behaviour" (J. R. Napier and P. H. Napier, eds.) 557–587. Academic
   Press, New York and London.
Groves, C. P. (1971). Geographic and individual variation in Bornean
   gibbons. *Folia primat.* **14**, 139–153.
Haeckel, E. (1866). "Generelle Morphologie der Organismen" Vol. 2.
   Georg Reimer, Berlin.
Hall-Craggs, E. C. B. (1965a). An osteometric study of the hind limb of
   the Galagidae, *J. Anat.* **99**, 119–125.
Hall-Craggs, E. C. B. (1965b). An analysis of the jump of the Lesser Galago
   (*Galago senegalensis*). *J. zool. Lond.* **147**, 20–29.
Hamlett, G. W. D. (1939). Reproduction in American monkeys. *Anat. Rec.*
   73, 171.
Hartman, C. G. (1931). The breeding season in monkeys, with special
   reference to *Pithecus (Macacus) rhesus. J. Mammal.* **12**, 129–142.
Hartman, C. G. (1932). Studies in the reproduction of the monkey *Maca-
   cus (Pithecus) rhesus*, with special reference to menstruation and
   pregnancy. *Contrib. Embryol.* **23**, 1–62.
Hassler, R. (1966). Comparative anatomy of the central visual systems in
   day- and night-active primates. *In* "Evolution of the Forebrain" (R.
   Hassler and H. Stephen, eds.) 419–434. G. Thieme Verlag, Stuttgart.

von Hayek, H. (1960). Die Lunge und Pleura. *In* "Primatologia" (H. Hofer, A. H. Schultz and D. Starck, eds.) Vol. 3, part 2, 588–625. Karger, Basel.

Heape, W. (1894). The menstruation of *Semnopithecus entellus*. *Phil. Trans. R. Soc. (B.).* **185**, 411–471.

Heape, W. (1896). The menstruation and ovulation of *Macacus thesus*. *Proc. R. Soc.* **60**, 202–205.

Hellman, M. (1928). Racial characters in human dentition. *Proc. Am. phil. Soc.* **67** 157–174.

Herlant, M. (1961). L'activité génitale chez la femelle *Galago senegalensis moholi* (Geoffroy) et ses rapports avec la persistance de phénomènes d'ovogénèse chez l'adulte. *Annls. Soc. r. zool. Belg.* 91, 1–15.

Hershkovitz, P. (1963). A systematic and zoogeographic account of the monkeys of the genus *Callicebus* (Cebidae) of the Amazonas and Orinoco River basins. *Mammalia* **27**, 1–80.

Hershkovitz, P. (1966a). Taxonomic notes on tamarins genus *Saguinus* (Callithricidae, Primates), with descriptions of four new forms. *Folia primat.* **4**, 381–395.

Hershkovitz, P. (1966b). On the identification of some marmosets family Callithricidae (Primates). *Mammalia* **30**, 327–332.

Hershkovitz, P. (1968). Metachromism or the principle of evolutionary change in mammalian tegumentary colors. *Evolution* **22**, 556–575.

Hershkovitz, P. (1970). Notes on Tertiary platyrrhine monkeys and description of a new genus from the late Miocene of Colombia. *Folia primat.* **12**, 1–37.

van Herwerden, M. A. (1905). "Bijdrage tot de Kennis van der Menstrueelen Cyclus en Puerperium" Akad proefschr. (Geneesk). E. J. Brill, Utrecht and Leiden.

van Herwerden, M. A. (1925). Some remarks on the polyoestrus of primates. *Anat. Rec.* **30**, 221–223.

Hill, J. P. (1932). The developmental history of the primates. *Phil. Trans. R. Soc. (B.)* **221**, 45–178.

Hill, J. P. (1965). On the placentation of *Tupaia*. *J. Zool.* **146**, 278–304.

Hill, J. P. and Burne, R. H. (1923). The foetal membrane and placentation of *Chiromys madagascariensis*. *Proc. zool. Soc. Lond.* **2**, 1145–1170.

Hill, W. C. O. (1936). On a hybrid leaf-monkey; with remarks on the breeding of leaf-monkeys in general. *Ceylon J. Sci. (B.)* **20**, 135–148.

Hill, W. C. O. (1948). Rhinoglyphics: epithelial sculpture of the mammalian rhinarium. *Proc. zool Soc. Lond.* **118**, 1–35.

Hill, W. C .O. (1949). Some points in the enteric anatomy of the great apes. *Proc. zool. Soc. Lond.* **119**, 19–32.

Hill, W. C. O. (1952). The external and visceral anatomy of the Olive Colobus monkey (*Procolobus verus*). *Proc. zool. Soc. Lond.* **122**, 127–186.

Hill, W. C. O. (1953). Caudal cutaneous specializations in *Tarsius*. *Proc. zool. Soc. Lond.* **123**, 17–26.

Hill, W. C. O. (1954). "Man's Ancestry". Heinemann, London.

Hill, W. C. O. (1955). "Primates: Comparative Anatomy and Taxonomy" Vol. 2 "Tarsioidea". Edinburgh University Press.

Hill, W. C. O. (1957). "Primates: Comparative Anatomy and Taxonomy" Vol 3. Edinburgh University Press.

Hill, W. C. O. (1958). Pharynx, oesphagus, stomach, small and large intestine. Form and position. In "Primatologia" (H. Hofer, A. H. Schultz and D. Starck, eds.) Vol. 3, part 1, 139–207. Karger, Basel.

Hill, W. C. O. (1959). The anatomy of Callimico goeldii (Thomas). Trans. Am. phil. Soc. 49, 1–116.

Hill, W. C. O. (1960). "Primates: Comparative Anatomy and Taxonomy" Vol. 4. Edinburgh University Press.

Hill, W. C. O. (1966). "Primates: Comparative Anatomy and Taxonomy" Vol. 6. Edinburgh University Press.

Hill, W. C. O. (1968). The genera of Old World Monkeys and Apes. In "Taxonomy and Phylogeny of Old World Primates with Reference to the Origin of Man" (A. B. Chiarelli, ed.). Rosenburg and Sellier, Turin.

Hill, W. C. O. (1969). Ontogeny of the gut pattern in Saimiri. Folia primat. 11, 166–174.

Hill, W. C. O. (1970). "Primates: Comparative Anatomy and Taxonomy" Vol. 8. Edinburgh University Press.

Hill, W. C. O. and Rewell, R. E. (1948). The caecum of primates; its appendages, mesenteries and blood supply. Trans. zool. Soc. Lond. 26, 199–256.

Hill, W. C. O., Porter, A. H. and Southwick, M. D. (1952). The natural history, endoparasites and pseudo-parasites of the tarsiers (Tarsius carbonarius) recently living in the Society's menagerie. Proc. zool. Soc. Lond. 122, 79–119.

Hingston, R. W. G. (1920). "A Naturalist in Himalaya". Witherby, London.

Hofer, H. (1954). Die cranio–cerebrale Topographic bei den Affen und ihre Bedeutung für die menschliche Schädelform. Homo. 5, 52–72.

Hofer, H. O. (1957). Zur Kenntnis der Kyphosen des Primaten-Schädels. Verh. anat. Ges. Frieburg 54–76.

Hofer, H. O. (1960). Studien zum Problem des Gestaltwandels des Schädels der Säugetiere, insbesondere der Primaten. Z. Morph. Anthrop. 50A, 299–316.

Hofer, H. O. (1969a). On the evolution of the cranio-cerebral topography in Primates. Ann. N.Y. Acad. Sci. 162, 15–24.

Hofer, H. O. (1969b). On the organon sublinguale in Callicebus (Primates. Platyrrhini). Folia primat. 11, 268–288.

Hofer, H. O. and Wilson, J. A. (1967). An endocranial cast of an early Oligocene Primate. Folia primat. 5, 148–152.

Hopwood, A. T. and Hollyfield, J. P. (1954). An annotated bibliography of the fossil mammals of Africa (1742–1950). In "Fossil Mammals of Africa" No. 8. British Museum (Natural History), London.

Howard, E. (1930). The X-zone of the suprarenal cortex in relation to gonadal maturation in monkeys and mice and the epiphyseal union in monkeys. *Anat. Rec.* **46**, 93–104.

Howell, F. C. (1967). Recent advances in human evolutionary studies. *Rev. Biol.* **42**, 471–513.

Hübrecht, A. A. W. (1899). Über die Entwickelung der Placenta von *Tarsius* und *Tupaia*. *Proc. 4th Int. Congr. Zool.* 343–412.

Huntington, G. S. (1913). The macroscopic anatomy of the salivary glands in the lower primates. *Stud. Cancer Columbia Univ.* **4**, 73–113.

Hürzeler, J. (1948). Zur Stammesgeschichte der Necrolemuriden. *Abh. schweiz. paläont. Ges.* **66**, 1–46.

Hürzeler, J. (1949). Neubeschreibung von *Oreopithecus bambolii* Gervais. *Abh. schweiz. paläont. Ges.* **66**, 1–20.

Hürzeler, J. (1954). Zur systematischen Stellung von *Oreopithecus*. *Verh. naturf. Ges. Basel*, **65**, 88–95.

Hürzeler, J. (1958). *Oreopithecus bambolii* Gervais. *Verh. naturf. Ges. Basel*, **69**, 1–48.

Hürzeler, J. (1968). Questions et réflexions sur l'histoire des anthropomorphes. *Ann. Paleont*, **54**, 195–233.

Ioannou, J. M. (1966). The oestrous cycle of the potto. *J. Reprod. Fertil.* **11**, 455–457.

Jennison, G. (1927). "Table of Gestation Periods and Number of Young" A. and C. Black, London.

Jewell, P. A. and Oates, J. F. (1969). Breeding activity in prosimians and small rodents in West Africa. *J. Reprod. Fertil. Suppl.* **6**, 23–28.

Jolly, C. J. (1967). Evolution of baboons. *In* "The Baboon in Medical Research" (H. Vagtborg, ed.) Vol. 2, 23–50. University of Texas Press, Austin.

Jolly, C. J. (1970). The large African monkeys as an adaptive array. *In* "Old World Monkeys: Evolution, Systematics and Behaviour" (J. R. Napier and P. H. Napier, eds.). Academic Press, New York and London.

Jones, R. T. (1967). The anatomical aspects of the baboon's wrist-joint. *S. Afr. J. Sci.* **63**, 291–296.

Kaiser, I. H. (1947). Absence of coiled arterioles in the endometrium of menstruating New World monkeys, *Anat. Rec.* **94**, 353–357.

Kälin, J. (1961). Sur les primates de l'Oligocène inférieur d'Egypte. *Ann. Palaeont*, **47**, 1–8.

Kawai, M. (1962). The birth season of the Japanese monkey. *Yaen.* **11**, 9–12 (in Japanese).

Kawai, M. and Mizuhara, H. (1959). An ecological study of the wild mountain gorilla (*G.g. beringei*). *Primates*, **2**, 1–42.

Keith, A. (1921). "Human Embryology and Morphology" 4th edition. Edward Arnold, London.

Kenneth, J. H. (1947). "Gestation Periods" 2nd edition, Techn. Comm. 5. Imp. Bur. Anim. Breedg. Genet.

Klaatsch, H. (1890). Uber den Descensus testiculorum. *Morph. Jb.* **16**, 587–646.

Klaatsch, H. (1892a). Uber embryonale Anlagen des Scrotums und der Labia majora bei Arctopitheken. *Morph. Jb.* **18**, 383.

Klaatsch, H. (1892b). Zur Morphologic des Mesenterialbildungen am Darmkannal der Wirbelthiere II Theil, Säugetiere. *Morph. Jb.* **18**, 609–716.

von Koenigswald, G. H. R. (1968). The phylogenetical position of the Hylobatinae. *In* "Taxonomy and Phylogeny of Old World Primates With References to the Origin of Man" (A. B. Chiarelli, ed.). Rosenberg and Sellier, Turin.

von Koenigswald, G. H. R. (1969). Miocene Cercopithecoidea and Oreopithecoidea. *In* "Fossil Vertebrates of Africa" (L. S. B. Leakey, ed.) Vol. 1. Academic Press, London and New York.

von Koenigswald, G. H. R. (1970). In press.

Kollman, M. and Papin, L. (1925). Études sur les lémuriens: anatomie comparée des fosses nasales et de leur annexes. *Archs. Morph. gén. exp.* **22**, 60 pp.

Kortlandt, A. (1962). Chimpanzees in the wild. *Scient. Am.* **206**, 128–138.

Kortlandt, A. (1966). Chimpanzee ecology and laboratory management. *Lab. Primate Newsl.* **5**, 1–11.

Kortlandt, A. (1968). Handgebrauch bei freilebenden Schimpansen. *In* "Handgebrauch und Verstandigung bei Affen und Frühmenschen" (B. Rensch, ed.). H. Huber, Berne and Stuttgart.

Kubota, K. and Iwamoto, M. (1967). Comparative anatomical and neurohistological observations on the tongue of slow loris (*Nycticebus concang.*) *Anat. Rec.* **158**, 163–175.

Kuehn, R. E., Jensen, G. D. and Morrill, R. K. (1965). Breeding *Macaca nemestrina*; a program of birth engineering. *Folia primat.* **3**, 251–262.

Kurtén, B. (1968). "Pleistocene Mammals of Europe". Weidenfeld and Nicolson, London.

Lampert, H. (1926). Zur Kenntnis der Platyrrhinen-Kehlkopfes. *Morph. Jb.* **55**, 607–654.

Lancaster, J. B. and Lee, R. B. (1965). The annual reproductive cycle in monkeys and apes. *In* "Primate Behaviour" (I. DeVore, ed.) 486–513. Holt, Rinehart and Winston, New York.

Lartet, E. (1856). Note sur un grand singe fossile qui se rattache au groupe des singes supérieures. *C. r. hebd. Séanc. Acad. Sci., Paris*, **43**, 210–222.

Leakey, L. S. B. (1962). See Bishop, W. W.: The mammalian fauna and geomorphological relations of the Napak volcanics, Karamoja. *Rec. Geol. Surv. Uganda.* 1–18.

Leakey, R. E. F. (1969). New Cercopithecidae from the Chemeron beds of Lake Baringo, Kenya. *In* "Fossil Vertebrates of Africa" (L. S. B. Leakey, ed.) Vol. 1. Academic Press, London and New York.

Lewis, G. E. (1933). Preliminary notice of a new genus of lemuroid from the Siwaliks. *Am. J. Sci.* **26**, 134–138.

van Loghem, J. J. (1903). Das Colon und mesocolon der Primaten. *Petrus Camper ned. Bijdr. Anat.* **2**, 350–437.

Lönnberg, E. (1937). On some fossil mammalian remains from East Africa. *Ark. Zool.* **29A**, 1–23.

Lowther, F. de L. (1940). A study of the activities of a pair of *Galago senegalensis moholi* in captivity, including the birth and post-natal development of twins. *Zoologica, N.Y.* **25**, 433–462.

Luckett, W. P. (1968). Morphogenesis of the placenta and foetal membrane of the tree shrews. *Am. J. Anat.* **123**, 385–427.

McCann, C. (1933). Notes on some of the Indian langurs. *J. Bombay nat. Hist. Soc.* **36**, 618–628.

McDowell, S. B. (1958). The Greater Antillean insectivores. *Bull. Am. Mus. nat. Hist.* **115**, 115–214.

McKenna, M. C. (1960). Fossil mammals of the early Wasatchian Four Mile fauna of northwest Colorado. *Univ. Calif. Publs. Bull. Dep. Geol.* **37**, 1–130.

McKenna, M. C. (1966). Paleontology and the origin of the primates. *Folia primat.* **4**, 1–25.

Manley, G. H. (1966). Reproduction in lorisoid primates. *Symp. zool. Soc. Lond.* **15**, 493–509.

Martin, R. D. (1966). Tree shrews: unique reproductive mechanism of systematic importance. *Science* **152**, 1402–1404.

Mason, W. A. (1966). Social organization of the south American monkey, *Callicebus moloch*; a preliminary report. *Tulane Stud. Zool.* **13**, 23–28.

Maxwell, M. (1928). The home of the eastern Gorilla. *J. Bombay nat. Hist. Soc.* **32**, 436–449.

Meisenheimer, J. (1921). "Geschlecht und Geschlechter im Tierreich". G. Fischer, Jena.

Meister, W. and Davis, D. D. (1956). Placentation of the pigmy tree-shrew *Tupaia minor. Fieldiana, Zool.* **35**, 73–84.

Meister, W. and Davis, D. D. (1958). Placentation of the terrestial treeshrew (*Tupaia tana*). *Anat. Rec.* **132**, 541–553.

Mijsberg, W. A. (1923). Uber den Bau des Urogenitalapparates bei den männlichen Primaten. *Verh. K. Akad. Wet.* **23**, (1), 92 pp.

Millot, J. (1952). La faune Malgache et le mythe Gondwanien. *Mém. Inst. scient. Madagascar* (A) **7**, 1–36.

Mitchell, P. C. (1905). On the intestinal tract of mammals. *Trans. zool. Soc. Lond.* **17**, 437–536.

Mitchell, P. C. (1916). Further observations on the intestinal tract of mammals. *Proc. zool. Soc. Lond.* 183.

Mivart, St. G. (1875). "Ape" *In* Encyclopedia Britannica. 9th edition.

Mizuhara, H. (1957). "The Japanese Monkey, its Social Structure" Sanichi-syobo, Kyoto. (In Japanese.)

Montagna, W. and Machida, H. (1966). The skin of Primates XXXII, The Philippine Tarsier (*Tarsius syrichta*). *Am. J. phys. Anthrop.* **25**, 71–75.

Moreau, R. E. (1952). Africa since the Mesozoic. *Proc. zool. Soc. Lond.* **121**, 869–913.

Moynihan, M. (1966). Communication in the titî monkey, *Callicebus*. *J. Zool. Lond.* **150**, 77–127.

Napier, J. R. (1970). Paleoecology and catarrhine evolution. *In* "Old World Monkeys: Evolution, Systematics and Behaviour" (J. R. Napier and P. H. Napier, eds.). Academic Press, New York and London.

Napier, J. R. and Davis, P. R. (1959). The fore-limb skeleton and associated remains of *Proconsul africanus*. *In* "Fossil Mammals of Africa" No. 16. British Museum (Natural History), London.

Napier, J. R. and Napier, P. H. (1967). "A Handbook of Living Primates". Academic Press, London and New York.

Narath, A. (1901). Der Bronchialbaum der Säugetiere und des Menschen. *Biblthca. med.* (A). **3**, 379 pp.

Negus, V. E. (1949). "The Comparative Anatomy and Physiology of the Larynx". William Heinemann, London.

Nissen, H. W. (1931). A field study of the chimpanzee. *Comp. Psychol. Monogr.* **8**, 1–122.

Noback, C. V. (1936). 1: Note on menstruation in the gorilla. 2: Note on gross changes observed in the external genitalia of the female gorilla just before, during and after menstruation. *Am. J. phys. Anthrop. Suppl.* **21**, 9.

Noback, C. V. (1939). Changes in the vaginal smears and associated cyclic phenomena in the lowland gorilla (*Gorilla gorilla*). *Anat. Rec.* **73**, 209–226.

Nusbaum, J. and Markowski, Z. (1896). Zur vergleichenden Anatomie der Stützorgane im der Zunge der Säugetiere. *Anat. Anz.* **12**, 551–561.

Nusbaum, J. and Markowski, Z. (1897). Weitere Studien über die vergleichenden Anatomie und Phylogenie der Zungenstützorgane der Saügetiere, Zugleich ein Beitrag zur Morphologie der Stutzgebilde im der menschen Zunge. *Anat. Anz.* **13**, 345–358.

Osborn, R. M. (1963). Behaviour of the Mountain Gorilla. *Symp. zool. Soc. Lond.* **10**, 29–37.

Ottow, B. (1955). "Biologische Anatomie der Genitalorgane und der Fortpflanzung der Säugetiere". G. Fischer, Jena.

Owen, R. (1833). On the stomachs of two species of *Semnopithecus*. *Proc. zool. Soc. Lond.* 74–76.

Owen, R. (1835). On the sacculated form of the stomach as it exists in the genus *Semopithecus* F. Cuv. *Trans. zool. Soc. Lond.* **1**, 65–70.

Owen, R. (1859). "Amphilestes". *In* Encyclopedia Britannica, 8th edition, Vol. 17, 157–158.

Patterson, B. (1954). The geologic history of non-hominid Primates of the Old World. *Hum. Biol.* **26**, 191–209.

Pehrson, T. (1914). Beiträge zur Kenntniss der äusseren Weiblichen Genitalen bei Affen, Halbaffen und Insectivoren. *Anat. Anz.* **46**, 161–179.

Percy, Professor (1844). On the management of various species of monkeys in confinement. *Proc. zool. Soc. Lond.* 81–84.

Petter, J. J. (1965). The lemurs of Madagascar. *In* "Primate Behaviour" (I. DeVore, ed.). Holt, Rinehart and Winston, New York.

Petter-Rousseau, A. (1962). Recherches sur la biologie de la réproduction des primates inférieurs. *Mammalia*, **26**, (Suppl. 1), 1–88.

Pilbeam, D. R. and Walker, A. C. (1968). Fossil monkeys from the Miocene of Napak, north-east Uganda. *Nature, Lond.* **220**, 657–660.

Pilgrim, G. E. (1915). New Siwalik Primates and their bearing on the question of the evolution of man and the Anthropoidea. *Rec. geol. Surv. India.* **45**, 1–74.

Pitman, C. R. S. (1935). The gorillas of the Kayonsa region. *Proc. zool. Soc. Lond.* 477–494.

Pocock, R. I. (1926). The external characters of the catarrhine monkeys and apes, *Proc. zool. Soc. Lond. for 1925.* 1479–1579.

Pohl, L. (1910). Beiträge zur Kenntnis des Os Penis der Prosimier. *Anat. Anz.* **37**, 225–231.

Polyak, S. (1957). "The Vertebrate Visual System". Chicago University Press.

Radinsky, L. B. (1967). The oldest primate endocast. *Am. J. phys. Anthrop.* **27**, 385–388.

Rahm, U. H. (1970). Ecology, zoogeography and systematics of some African forest monkeys. *In* "Old World Monkeys: Evolution, Systematics and Behaviour" (J. R. Napier and P. H. Napier, eds.) 591–626. Academic Press, New York and London.

Raven, H. C. (1936). Genital swelling in a female gorilla. *J. Mammal.* **17**, 416.

Rengger, J. R. (1830). "Naturgeschichte der Säugetiere von Paraguay". Schweighauserchen Buchhandlung, Basel.

Reynolds, V. (1965). "Budongo, a Forest and its Chimpanzees". Methuen, London.

Reynolds, V. and Reynolds, F. (1955). Chimpanzees of the Budongo forest. *In* "Primate Behaviour" (I. DeVore, ed.). Holt, Rinehart and Winston, New York

Riethe, P. (1967). Zur formale Wandlung des Dryopithecus-musters. *Dt. zahnärtzl. Ztg.* **22**, 819–823.

Robinson, J. T. and Allin, E. F. (1966). On the Y of the Dryopithecus pattern of mandibular molar teeth. *Am. J. phys. Anthrop.* **25**, 323–324.

Rowell, T. E. (1970). Reproductive cycles of two *Cercopithecus* monkeys. *J. Reprod. Fertil.* **22**, 321–338.

Saayman, G. S. (1970). The menstrual cycle and sexual behaviour in a troop of free ranging chacma baboons (*Papio ursinus*). *Folia primat.* **12**, 81–110.

Sabater Pi, J. (1964). Distribution actual de los Gorilas de Llanura en Rio Muni. *Publ. Serv. Municip. Parque zool.* Barcelona.

Sabater Pi, J. (1970). Apostación a la ecologia de los *Colobus polykomos satanas* Waterh. de Rio Muni. (In press).

Sade, D. S. (1964). Seasonal cycle in size of testes of free-ranging *Macaca mulatta. Folia primat.* **2**, 171–180.

Schaffer, J. (1940). "Die Hautdrüsenorgane der Säugetiere". Urban and schwarzenburg, Berlin and Vienna.

Schaller, G. B. (1963). "The Mountain Gorilla: Ecology and Behaviour". Chicago University Press.

Schaller, G. B. (1965). The behaviour of the Mountain Gorilla. In "Primate Behaviour" (I. DeVore, ed.). Holt, Rinehart and Winston, New York.

Schneider, R. (1958). Zunge und weicher Gaumen. In "Primatologia" (H. Hofer, A. H. Schultz and D. Starck, eds.) Vol. 3, part 1. Karger, Basel.

Schultz, A. H. (1938a). Genital swelling in the female orang-utan. J. Mammal. **19**, 363–366.

Schultz, A. H. (1938b). The relative weight of the testes in Primates. Anat. Rec. **72**, 387–394.

Schwalbe, G. (1916). Uber den fossilen Affen Oreopithecus bambolii. Z. Morph. Anthrop. **19**, 149–254, 501–504.

Sera, G. L. (1935). I caratteri morfologici di Palaeopropithecus e l'adattamento acquatico primitive dei mammiferi e dei Primati in particolari. Arch. ital. Anat. Embriol. **35**, 229–270.

Sera, G. L. (1938). Alcuni caratteri scheletri di importanza ecologica e filetica nei Lemuri fossili ed attuali. Palaeontogr. ital. **38**, 1–112.

Sera, G. L. (1947). L'occultamento subacqueo e l'ancoramento nella paleobiologia di parte del monodelfi e la distenzione di questi in due gruppi filetici. Palaeontogr. ital. **41**, 63–120.

Sera, G. L. (1950). Ulterior osservazioni sui Lemuri fossili ed attuali. Palaeontogr. ital. **47**, 1–97.

Seth, P. K. (1964). The crista sagittalis in relation to the nuchal crest in Nycticebus coucang. Am. J. phys. Anthrop. **22**, 53–64.

Simons, E. L. (1961a). Notes on Eocene tarsioids and a revision of some Necrolemurinae. Bull. Br. Mus. nat. Hist. (Geol.). **5**, 45–69.

Simons, E. L. (1961b). The phyletic position of Ramapithecus. Postilla. **57**, 1–9.

Simons, E. L. (1962). A new Eocene primate Cantius, and a revision of early Cenozoic lemuroids of Europe. Bull. Br. Mus. nat. Hist. (Geol.). **7**, 1–36.

Simons, E. L. (1963). A critical reappraisal of Tertiary Primates. In "Evolutionary and Genetic Biology of Primates" (J. Buettner-Janusch, ed.) Vol. 1, 65–129. Academic Press, New York and London.

Simons, E. L. (1964). The early relatives of man. Scient. Am. **211**, 14 pp.

Simons, E. L. (1965). New fossil apes from Egypt and the initial differentiation of Hominoidea. Nature, Lond. **205**, 135–139.

Simons, E. L. (1970). The deployment and history of Old World monkeys (Cercopithecoidea, Primates). In "Old World Monkeys: Evolution, Systematics and Behaviour" (J. R. Napier and P. H. Napier, eds.). Academic Press, New York and London.

Simons, E. L. and Pilbeam, D. R. (1965). Preliminary revision of the Dryopithecinae (Pongidae, Anthropoidea). Folia primat. **3**, 81–152.

Simpson, G. G. (1940). Studies in the earliest Primates. *Bull. Am. Mus. nat. Hist.* **77**, 185–212.

Simpson, G. G. (1945). The principles of classification and a classification of mammals. *Bull. Am. Mus. nat. Hist.* **85**, xvi + 350 pp.

Simpson G. G. (1955). The Phenacolemuridae, new family of early Primates. *Bull. Am. Mus. nat. Hist.* **105**, 412–442.

Simpson, G. G. (1963). The meaning of taxonomic statements. In "Classification and Human Evolution" (S. L. Washburn, ed.). Aldine Publishing Company, Chicago.

Simpson, G. G. (1967). The Tertiary lorisiform Primates of Africa. *Bull. Mus. comp. Zool. Harv.* **136**, 39–61.

Sonntag, C. F. (1921). The comparative anatomy of the tongues of the mammalia. *Proc. zool. Soc. Lond.* 1–29, 277–322, 497–524, 741–755, 757–767.

Sonntag, C. F. (1924). "The Morphology and Evolution of the Apes and Man". John Bale, Sons and Danielsson, London.

Sorenson, M. W. and Conaway, C. H. (1964). Observations on tree-shrews in captivity. *Sabah Soc. J.* **2**, 77–91.

Sorenson, M. W. and Conaway, C. H. (1968). The social and reproductive behaviour of *Tupaia montana* in captivity. *J. Mammal.* **49**, 502–512.

Southwick, C. H., Beg, M. A. and Siddiqui, M. R. (1961). A population survey of Rhesus monkeys in villages, towns and temples in northern India. *Ecology* **42**, 538–547.

Sprankel, H. (1961). Histologie und biologische Bedeutung eines jugulosternalen Duftdrüsenfeldes bei *Tupaia glis* Diard 1820 (*Tupaia* scent marking). *Verh. dt. zool. Ges.* 198–206.

Sprankel, H. (1965). Untersuchungen an *Tarsius* 1, Morphologie des Schwanzes nebst ethologischen Bemerkungen. *Folia primat.* **3**, 153–188.

Sprankel, H. (1970). Zur vergleichenden Histologie von Hautdrusenorgane in Lippenbereich bei *Tarsius bancanus* und *Tarsius syrichta*. Abstract 3rd International Congress of Primatology, Zürich. p. 58.

Stanley, W. B. (1919). Carnivorous apes in Sierra Leone. *Sierra Leone Stud.* 3–19.

Starck, D. and Schneider, R. (1960). Respirationsorgane. In "Primatologia" (H. Hofer, A. H. Schultz and D. Starck, eds.) Vol. 3, part 2, 423–587. Karger, Basel.

Stehlin, H. G. (1916). Die Säugetiere des Schweizerischen Eocäens. *Abh. schweiz-paläont. Ges.* **41**, 1299–1552.

Stirton, R. A. and Savage, D. E. (1951). A new monkey from the La Venta late Miocene of Colombia. *Minis. Min. Pe. Serv. Geol. Nac.* **7**, 347–356.

Strahl, H. (1899). Der Uterus gravidus von *Galago agisymbanus*. *Abh. senckenb. naturforsch. Ges.* **XXVI**, 155–199.

Straus, W. L. (1931). The form of the tracheal cartilages in Primates. *J. Mammal.* **12**, 281–285.

Straus, W. L. (1963). The classification of *Oreopithecus*. *In* "Classification and Human Evolution" (S. L. Washburn, ed.). Aldine Publishing Company, Chicago.

Struhsaker, T. T. (1967). Behaviour of vervet monkeys and other *Cercopithecus*. *Science* **156**, 1197–1203.

Swinnerton, H. H. (1960). "Fossils" (New Naturalist No. 42). Collins, London.

Szalay, F. S. (1968). The beginnings of Primates. *Evolution* **22**, 19–36.

Tappen, N. C. (1960). Problems of distribution and adaptation of the African monkeys. *Curr. Anthrop.* **1**, 91–118.

Tattersall, I. (1968). A mandible of *Indraloris* (Primates Lorisidae) from the Miocene of India. *Postilla* **123**, 1–10.

Thenius, E. (1954). Die Bedeutung von *Austriacopithecus* Ehrenberg für die Stammesgeschichte der Hominoidea. *Anz. öst. Akad. Wiss.* **13**, 191–196.

Tobias, P. V. (1968). Taxonomy and phylogeny of the Australopithecines. *In* "Taxonomy and Phylogeny of Old World Primates With References to the Origin of Man" (A. B. Chiarelli, ed.). Rosenberg and Sellier, Turin.

Tokuda, K. (1962). A study of the sexual behaviour in the Japanese monkey troop. *Primates*. **3**, 1–40.

Tomilin, M. I. (1940). Menstrual bleeding and genital swelling in *Miopithecus talapoin*. *Proc. zool. Soc. Lond.* **110A**, 43–45.

Turner, W. (1876). On the placentation of the lemurs. *Phil. Trans. 1876* **166**, 569–587. *J. Anat.* **12**, 147–153.

van Valen, L. (1965). Treeshrews, Primates and fossils. *Evolution* **19**, 137–151.

Verma, K. (1965). Notes on the biology and anatomy of the Indian tree-shrews, *Anathana wroughtoni*. *Mammalia*, **29**, 289–330.

Vogel, C. (1962). Untersuchungen an *Colobus*-Schädeln aus Liberia unter besonderer Berucksichtigung der Crista sagittalis. *Z. Morph. Anthrop.* **52**, 306–332.

Vogel, C. (1968). The phylogenetical evolution of some characters and some morphological trends in the evaluation of the skull in catarrhine primates. *In* "Taxonomy and Phylogeny of Old World Primates With References to the Origin of Man" (A. B. Chiarelli, ed.) 21–56. Rosenberg and Sellier, Turin.

van Wagenen, G. (1945). Optimal mating time for pregnancy in the monkey. *Endocrinology*, **40**, 307–312.

Walker, A. C. (1967). Patterns of extinction among the subfossil Madagascan lemuroids. *In* "Pleistocene Extinctions" (P. S. Martin and H. E. Wright, eds.). Yale University Press.

Walker, A. C. (1969). True affinities of *Propotto leakeyi*. *Nature, Lond.* **223**, 647–648.

Walker, A. C. (1970). Post-cranial remains of the Miocene Lorisidae of East Africa. *Am. J. phys. Anthrop.* **33**, 249–261.

Washburn, S. L. (1950). Thoracic viscera of the gorilla. *In* "The Anatomy of the Gorilla" (W. K. Gregory, ed.) Raven Memorial Volume. Columbia University Press, New York.

Washburn, S. L. (1957). Ischial callosities as sleeping adaptations. *Am. J. phys. Anthrop.* **15**, 269–276.

Watson, D. M. S. (1942). On Permian and Triassic tetrapods. *Geol. Mag.* **79**, 81–116.

Weber, M. (1928). "Die Säugetiere; Systematischer Teil". G. Fischer, Jena.

Wegner, R. N. (1956). Studien über Nebenhöhlen des Schadels. *Wiss. Z. Univ. Greifswald*, **1**, 55 pp.

Weinert, H. (1925). Die Ausbildung der Stirnhöhlen als Stammgeschichtliches Merkmal. *Z. Morph. Anthrop.* **25**, 243–357, 365–418.

Wharton, C. H. (1950). Notes on the Philippine tree shrew *Urogale everetti* (Thomas, 1892). *J. Mammal.* **31**, 352–354.

Wilson, J. A. (1966). A new Primate from the earliest Oligocene, West Texas; preliminary report. *Folia primat.* **4**, 228–248.

Williams, E. F. and Koopman, K. F. (1952). West Indian fossil monkeys. *Amer. Mus. Novit.* **1546**, 16 pp.

Winge, H. (1924). "Pattedyr-Slaegter". H. Hagerups Forlag, Copenhagen.

Wislocki, G. B. (1929). On the placentation of the primates, with a consideration of the phylogeny of the placenta. *Contrib. Embryol.* **20**, 51–80.

Wislocki, G. B. (1930). On a series of placental stages of a platyrrhine monkey (*Ateles geoffroyi*) with some remarks upon age, sex and breeding period in platyrrhines. *Contrib. Embryol.* **22**, 173–192.

Wislocki, G. B. (1932). On the female reproductive tract of the gorilla with a comparison with that of other primates. *Contrib. Embryol.* **23**, 163–204.

Wislocki, G. B. (1933). The reproductive systems. *In* "The Anatomy of the Rhesus Monkey" (C. G. Hartman and W. L. Straus, eds.). Baillière, Tindall and Cox, London.

Wislocki, G. B. (1936). The external genitalia of simian Primates. *Hum. Biol.* **8**, 309–347.

Wislocki, G. B. and Straus, W. L. (1932). On the blood-vascular bundles in the limbs of certain edentates and lemurs. *Bull. Mus. comp. Zool. Harv.* **74**, 1–15.

Wood Jones, F. (1916). "Arboreal Man". Edward Arnold, London.

Wood Jones, F. (1917). The genitalia of *Tupaia*. *J. Anat.* **51**, 118–126.

Wood Jones, F. (1923). "The Mammals of South Australia". Government Printer, Adelaide.

Wood Jones, F. (1929). "Man's Place among the Mammals". Edward Arnold, London.

Wood Jones, F. (1939). The so-called maxillary antrum of the gorilla. *J. Anat.* **73**, 116–119.

Woollard, H. H. (1942). The cortical lamination of *Tarsius*. *J. Anat.* **60**, 86–105.

Yerkes, R. M. and Yerkes, A. (1929). "The Great Apes: A Study of Anthropoid Life". Yale University Press.

Zapfe, H. (1952). Die *Pliopithecus* Funde aus der Spaltenfüllung von Neudorf an der March (C.S.R.). *Sonderh. geol. Bundesanst.* **C**, 1–5.

Zapfe, H. (1961). Ein Primatenfund aus der Miozänen Molasse von Oberösterreich. *Veröff. naturh. Mus. Wien.* **1**, 1–5.

Zuckerman, S. (1930). The menstrual cycle of the primates—Part 1, General nature and homology. *Proc. zool. Soc. Lond.* **2**, 691–754.

Zuckerman, S. (1931). The alleged breeding season of primates, with special reference to the Chacma Baboon (*Papio porcarius*). *Proc. zool. Soc. Lond.* **1**, 325–343.

Zuckerman, S. (1932a). The comparative physiology of the menstrual cycle, *Brit. Med. J.* **2**, 1093–1097.

Zuckerman, S. (1932b). Further observations on the breeding of primates, with special reference to the suborders Lemuroidea and Tarsioidea. *Proc. zool. Soc. Lond.* 1059–1075.

Zuckerman, S. (1937). The duration and phases of the menstrual cycle in primates. *Proc. zool. Soc. Lond.* 315–329.

Author Index

Numbers in *italics* indicate the pages on which the references are listed.

Subject Index